DOMESTIC PARTICULARS

ALSO BY FREDERICK BUSCH

Fiction

Breathing Trouble (stories)
I Wanted a Year without Fall (novel)
Manual Labor (novel)

Criticism

Hawkes: A Guide to His Fictions

frederick busch
DOMESTIC PARTICULARS

A Family Chronicle

A NEW DIRECTIONS BOOK

Copyright © 1972, 1973, 1974, 1975, 1976 by Frederick Busch

All rights reserved. Except for brief passages quoted in a newspaper, magazine, radio, or television review, no part of this book may be reproduced in any form or by any means, electronic or mechanical, including photocopying and recording, or by any information storage and retrieval system, without permission in writing from the Publisher.

Portions of this book originally appeared, sometimes in substantially altered form, in *The Carleton Miscellany, New Directions in Prose and Poetry* (nos. 28 and 31), and *Penguin Modern Stories 9*. "How the Indians Come Home" and "The Trouble with Being Food" were first published in *Esquire*. The author is grateful to the editors and publishers of these books and magazines, and to the Colgate University Research Council for its generous support.

Grateful acknowledgment is also made to the editors and publishers of books where some of the material quoted in this volume first appeared and from whom permission to reprint has been obtained: Farrar, Straus & Giroux, for portions of "He Resigns," from *Delusions, Etc.* by John Berryman (Copyright © 1969, 1971 by John Berryman; Copyright © 1972 by the Estate of John Berryman); Harcourt Brace Jovanovich, for a passage from Quentin Bell's *Virginia Wolf: A Biography* (Copyright © 1972 by Quentin Bell); Harper & Row, for portions of "Lady Lazarus," from *Ariel* by Sylvia Plath (Copyright © 1963 by Ted Hughes); Random House, for a passage from Sigmund Freud's *The Interpretation of Dreams* (Copyright 1950 by Random House, Inc.).

Manufactured in the United States of America
First published clothbound and as New Directions Paperbook 413 in 1976
Published simultaneously in Canada by McClelland & Stewart, Ltd.

Library of Congress Cataloging in Publication Data

Busch, Frederick, 1941–
    Domestic particulars.

    (A New Directions Book)
    I. Title.
PZ4.B9767Do   [PS3552.U814]   813'.5'4   76–8904
ISBN 0–8112–0605–X
ISBN 0–8112–0611–4 pbk.

New Directions Books are published for James Laughlin
by New Directions Publishing Corporation,
333 Sixth Avenue, New York 10014

For Nick

DOMESTIC PARTICULARS

A THREE-LEGGED RACE
1919–1951

I had something terrible in my eye. It turned red, then the lid lowered until I looked like a pirate. I was sure that's how I looked though I had only heard about pirates, never seen a picture of one. In 1919, orthodox Jewish girls didn't look at books about pirates. My mother took me to the drugstore. She wore no coat although it was autumn and cold in Manhattan. She made me wear my older sister's handed-down coat. To shame me with her martyrdom? Maybe only because I hated my sister, who made my life terrible. A punishment of warmth.

We couldn't afford to consult the druggist, but we could afford a doctor less. She told me this. So I felt wretched, the cause of more poverty, as we left. Because it was Saturday, all the lights in the apartment were off. The foyer door from our building opened out and wonderful light came in. I was surprised that the street looked so happy. She held my hand, but not with affection. She showed her affection only at night, when she was tired and would have to live until morning with her memories of too little food for us, clothing made of rags, our father's furious noises, all the crying. Then, as we went to bed, she told us stories of Austria and in Yiddish said wisdoms. All she could give. But on the street she held my hand to keep me. As a cat coldly carries its young in its mouth. Efficiency. We didn't talk.

Past the Yiddish theater, the newspaper kiosk, across the trolley tracks we couldn't ride on Saturday, the kosher shops which were closed because on the Sabbath money was not to be touched. Bessermann, a Jew but more a merchant, in his store that smelled of soap and ginger and vanilla, was like a king. Fat-bellied but small in the face, with little eyes and little fingers, a mouth that also smelled of soap, he held my face and turned it, pulled my eyelid up, made me roll my eyeball and weep.

"Tessie," he told my mother, dropping my face from his hand, "you'll leave the eye here overnight, I'll have it ready in maybe two days. Yes?"

I said nothing. I wept harder but I said nothing. Then he folded a shape of blue paper, poured some powder in, told my mother how to use it. She never smiled or spoke, except to tell him that on Monday he'd be paid. He nodded and forgot me. And in the street I howled with horror. My mother took my hand and pulled me home. At night she spoke to me of jokes and teasing. But she left me in her silence during the rest of the long afternoon.

I wonder if I believed him when he said it. I have believed so much. I wonder now, at the middle of the century I started in, whether I was really afraid. How much was I disappointed? That the torture, theft, bereavement, heroism, *magic* didn't take place? Bessermann was only a human male. This is the problem of modern man. My problem too.

Weakness. As when I was working for Liveright, taking dictation, typing, even reading some manuscripts. People coming in drunk with cocktails in their hands. The long talks about music and drama. The way they would remember finding Hemingway's stories with their little italicized chapters between a book about African sex and a manuscript on Marxist art criticism. They weren't doing well, then, and I earned very little. But it was the most exciting time of my life. I was alive, in touch with living people and everything important in the world. Sometimes I didn't eat lunch for fear of missing an irreplaceable accidental event.

With this there was Alvin, finishing his residency at St. Vincent's, eating meals in Greenwich Village, where I lived on 11th Street, the most exciting place in the world besides Paris. We would meet sometimes and talk of dreams. He wanted to practice medicine on an island in the Pacific. We wanted everything simple, fundamental. Life could be *touched,* we said. He would heal people and I would write a book about us. We believed that. We believed in Edna St. Vincent Millay, the unbitter parts of her poems. What we wanted, I still believe in that. You could have it now if you wanted to deserve it. Even now.

But his Jew-bitch mother had the soul of an adding machine. I never met her. He never brought me home. She told him that an office with a nurse and a large enough car would cost him $10,000, which he didn't have. Only a girl with money, easily married and jailed at the back of the private house she bought him. Only that. Not me. So Alvin wrote me a letter and explained. I have never accepted explanations since then.

I continued to live on 11th Street, meeting interesting people, seeing plays, refusing to write to my mother. My father by then had left her. A drunken baker addicted to cough syrup. I called myself an orphan. I was dreaming alone. I met Mrs. Miriam and visited often in her wonderful flat on Washington Square. She was elegant. I never thought of her as a mother, of course. I had no mother. But she taught me how to cook, and when I didn't work in the office on Saturdays, she let me sell the jewelry she made in the shop she ran on Grove Street. I loved handling the money, and sometimes, when no customers were there, I rang NO SALE and scooped the dimes and nickels into my hands, let them run like water back down. Her husband was an alcoholic. Weakling. But a very handsome man. He was always asleep when I came. She made him keep out of the way. She introduced me to Eugene O'Neill. Now I talk to janitors and pension clerks. She's old. I don't know where she is. She's ancient. I'm afraid to see her or speak on the phone. There is nothing to say. I wonder how she stays alive. She's alone now. Who isn't?

I married Mac because he was more of a virgin than I was. He was afraid to take me to his mother too. He wouldn't sleep in my apartment until we were married. Then he moved in. I had assumed he would find a larger place for us to live in. But he let me do the finding. Always. The new apartment in Brooklyn. Then the Brooklyn house. He loved me more than I loved him. He joined the Party because I did. Quit when I quit. He followed me everywhere, even in the house. I wasn't surprised. What surprised me was that I needed him. The day after we were married and had finally made love, after a night when he wouldn't let me speak of Alvin any more, I was standing at the door of the apartment with my coat on. From the day bed he said "Where are you going?" I told him out for a walk. He said "What about me?" I told him I'd forgotten. I waited for him to dress.

I had two miscarriages before Harry. They told me I might die if we tried again. Mac said no. But I said yes. We tried again. It almost killed me, but it didn't, and there was my child. There. *Where?*

And when Mac was in the army during the war, Harry had pneumonia. We didn't have the money for sulfa drugs, but I got the money. I always did what I had to. For three nights, I tied myself with Venetian blind cords to the slats of his crib so I would stay awake to take care of him. In the morning Mac's mother would come, with a look on her face as if it was my fault I had to go to work to make the money to keep him alive. Not that she would give me a dime. And at night I would come home and eat standing up beside the crib. And then the ropes, and watching him all night. He lived, and there was my child.

At thirty-two a nearsighted teacher had to join the army. It happened in the movies that way. I'm sure that's where he got the idea. He loved the movies. All that glory and heroism. They always lied, the heroes. They did anything to get in and get shot. Like a man in the movies, he memorized the eye charts, and they took him. So he could leave us behind, I thought. Or wondered. I wondered about that. I had a right to. Where were our dreams of long walks and long talks? Of living an intelligent life of excitement? In the army, being shot. Every day he wrote about how much he missed us. Training in Colorado, then Texas, then going overseas. Telling me. I remember it because it was so stupid and he said it all the time: *I won't get killed. They can't kill me. I'll come home safe.* And I was always surprised, when I wasn't panicked about money or Harry's health, to learn how much I needed Mac to live and return.

So today there's a new disease. Anorexia nervosa. Young girls get it. They starve. They vomit when they eat, even if they're hungry. Once it was polio and pneumonia, now this modern sickness. The girls are the ones who run away from home. Promiscuous, taking drugs. The kind of girl who lived with that killer Manson, the sex maniac. Doing anything with anyone. This is the age of disgust. They look like Auschwitz prisoners toward the end. Living skeletons. And an article in the *Times* blames it on whom? The mother. Of course. I don't remember. Either the mother feeds them too much, says "Eat, eat!" like the Jewish mother everyone makes fun of, our culture's new scapegoat, or she takes her breast away. It threatens them. So how can a mother win, with such "research"? She feeds them or she doesn't, and either way she's a monster. *That's* the disgusting part. Thank God I didn't have a daughter.

I had a son. My mouth used to water when I made him mashed potatoes with milk and butter mixed in. Now he hates mashed potatoes. He had a piano recital in elementary school. I asked him what he wanted me to wear. He told me the plaid dress with the high collar. Made of wool. In June. I wore it. I sweated for an hour to hear him play "I have a little Scottie dog/A little Scottie dog have I." Now he doesn't touch an instrument. Go be a mother.

I have studied at Columbia, N.Y.U., Rutgers, and a summer at Harvard, not to mention the A.B. from Hunter, and nothing has prepared me for this. Brooklyn, where we used to live in

Flatbush, was green. No cars. Empty lots with trees and bushes in them. The children played in the street and not too many cars came. It was middle class, then upper middle class, then they built the yellow-brick yeshivah on the corner. Then the people across the street, all with red hair and loud voices, the kind who get Jews persecuted, cut their two old maple trees down to build a patio on the front lawn with a high wooden fence for privacy when they gave their parties. Soon the Negroes came, and that was that. Mac got the job at Columbia, we left. I was afraid to stay. Not because of Negroes. Because of *no values*. Nobody respected anything I had learned to respect. Sidewalks, buildings, shrubbery, physical human bodies. Everything was endangered. It was time to go.

Here it isn't much better. People should learn not to dream.

I dieted for months before Mac came home from the service, healed from his wound. I look at pictures of us. I have hundreds and hundreds, maybe a thousand. Pictures. We look ancient. The skirts high, but a childish height. The hair of the women piling over the forehead, or onto the nape, or over the ear. Nothing balanced—*cocky* is the word. No matter how frightened, there was jauntiness in our faces and even our feet. Not like the English. They went crazy when Hitler said he would invade them, and they never recovered. Every morning, ten years later, they tighten their belts in case they're attacked. No. But we were sure. There was Hitler versus us. There was evil, there was good. We were good. We would win. We won, so the A-bomb was good. The Japanese were bad, so the A-bomb killed them and they lost. They were like Negroes to us. Oilier, more elusive. Like Negroes who had forgotten subservience. Pearl Harbor was an early example of Black Power. Only later on we couldn't drop the A-bomb on Harlem in retaliation without killing the landlords and policemen, who then were mostly white.

But we didn't *think* about that. We thought about us versus them and good over bad and we won. In the movies, when I could go, which wasn't too often, I cried. When the uneducated and undereducated wept, so did I. For Myrna Loy and Jean

Arthur and Raymond Massey. William Bendix. My God, for William Bendix: the stupid feeling heart of us all. Lloyd Nolan! Because to me in that time of my life, when I might have been half insane, but not knowing it, life unrolled simply before me. No matter how horrible or hard. It nevertheless was clear. It was a road I went down.

So Mac lived through the war. I did. Harry did. I dieted forever and was nearly thin on the day I was to take the BMT to 42nd Street, change for the shuttle, go to Grand Central, and meet his train from the West Coast. I walked to the subway station. On the steel and concrete bridge over the Long Island Railroad freight line tracks on Avenue H, trees blowing and the sky as blue as a summer snapshot, the tall handsome soldier walked past me. I smiled and so did he. My stomach tightened, I thought of my body as provocative. I thought of *men*. From the other end of the bridge he called me: "Claire! Claire!" Our reunion.

But I recognized him after that. And he knew me. He got his job back at Brooklyn College, though they made him teach freshman composition for years during the slump. I worked at my degree during summers and nights. During the weekdays I worked for Henry Holt. I free-lanced for Harcourt. In Manhattan I was competent. At home in Brooklyn I made meals and interviewed maids. One stole my wallet during the interview. One, a German, Mac felt sorry for her, lived in. Every Friday night I made us go out. To see publishing friends, never his colleagues. I never met his colleagues except by accident. We came home early one night to see her drunk with two men. Two clients. With Harry sitting on the hall steps, listening. So I have been a kind of madam too. I studied and I went to school. I taught part-time at night on an assistantship. I nearly went crazy. I won my Ph.D. But no one hired Jewish ladies in English departments. I edited books for men who were editors.

Mac in his bed dreamed of death. He rolled, he bit his lip, he cried and whimpered. I woke him when the nightmares came and asked him to tell me what it was. He said "The war." For years he told me nothing. It was his. I had my bed and my own nightmares. We met in the morning. Two pieces of toast with peanut butter and marmalade. Two cups of coffee. Every morning he ate that meal and on the way to school, he never told me where, had another cup of coffee. He read Freud for help in analyzing his dreams, and they stopped. For him it was always that simple. My dreams went on in the daytime.

He was promoted, and I was given raises, though never better jobs. I stopped applying to teaching posts and taught myself Russian. Everyone said the Russians would conquer the world. Maybe they would let me teach. Then post cards came from my mother. She lived on the Lower East Side, still. "Dear daughter" they began. In the middle they always said "I know you will try to send what you can." At the end, "I hope Mac and the children are well." I never told her Harry was all that we had. Mac sent her money at the end of every month. And then my father sued us, all the children, for support. We sent nothing. Mac went to court. I couldn't. I couldn't. We didn't hear from him again until my youngest sister wrote me that he died in a county hospital. I didn't tell Mac for a month.

Mac's parents lived with his sister, Ida. She and Dick drove them back to New York from Monticello where they were sick. His mother hated me, always. She blamed me when he went to the war. She blamed me whenever Harry was sick. When Mac was in the hospital for his back, she told me I didn't take care of him. They didn't know how sick she was. She died bleeding from the vagina. It was horrible pain and I hated them —his sister and his brother and him. Letting her die. Alone. Yes, in a room where her daughter lived. Yes, visited by her children. Yes, attended by the weak old man, her husband, who in Russia was an anarchist and spat at rabbis, quailing now and flimsy and full of fear. But alone.

I made the arrangements for the wake. He was strong for his brother and sister, and he held his father as if his father were his child. Then he wept and coughed when we were alone in the house. He let me hold him in our living room and on the stairs. When we pulled Harry's covers up on his shoulders. In our bedroom beside my bed. Then he told me his heart was pounding. He shook and sweated, felt his own pulse. He told me that his chest hurt. He couldn't breathe. He lay in his bed and I watched him till he slept. We never went to bed together again.

We went to the nineteen-fifties. Arguments about Eisenhower. Mac thought he looked like Ike and voted for him, though his colleagues all voted for Stevenson. I didn't vote. Everyone we'd seen in leftist theater in the thirties was turning everyone else in to congressional committees. We got Harry a TV set, and the programs were about Communists trying to subvert the American way. The TV went in the semifinished basement (Mac painted it), near the furnace. I let Harry watch an hour a day during the week, an hour and a half on Saturday, two hours on Sunday. Mac watched the "Colgate Comedy Hour" with Harry and howled at Jerry Lewis until the tears ran on his face. I read upstairs. I liked Mister Peepers: he reminded me of me. Mac said there was no resemblance. And at Brooklyn College, the left-wing teachers were purged. Mac kept his mouth shut and survived.

We had pretty much stopped going to parties by then. We sometimes went to museums or on a Sunday field trip, but Harry hated that and I hated his sulking. We stayed home. They did. I went to the city by myself as often as I could. I wanted to go on field trips with the Brooklyn Bird Club by myself. But Mac insisted on coming. He never initiated. He always went along. Which meant Harry had to come, because who can you trust with a pre-adolescent all day long? They were hateful days, and I was happy to come home so they would watch their Sunday television.

When I complained that Mac never took me to lunch near the campus, or had his colleagues over, he planned a party. All I had to do was complain. Then he'd act. He invited six army buddies and their wives or girlfriends for a Friday night. I came home from work in the early afternoon and cooked. A *crème caramel,* hot beef hors d'oeuvres in cream, veal birds stuffed with ham, fresh broccoli, bottles of liquor and soda, freshly ironed tablecloths and napkins. Harry wore a sport coat we had bought because Mac said a boy of eleven should own a sport coat. He whined when I made him put it on, but then he got excited. He got ready to be cute and winning for the guests, I could see that.

They came in their business suits and party dresses. They had never met me. Mac had never brought us together after the war. We sat in a ring in the living room and I made each one a highball, measuring the ounce and a half of liquor in a silver cup, two ice cubes per drink so we wouldn't run out of ice. Mac wasn't home by six, and then he wasn't home by seven. At nine they were itchy with embarrassment for me. Harry chattered and amused them. Then he bored them. We ate. I made more drinks for them. I wouldn't drink anything. I pressed my hands in my lap and waited. At ten-fifteen the phone rang and they watched me answer it in the hall. It was Mac, so drunk he hardly could talk. He said "Claire? Claire? I'm at a corner. Can you see me? What's the corner called, Claire? I don't know where I am. Do *you* know where I am?" I made him describe the street and I told him to stay there. Then I called the police and asked them to find him.

So I met Mac's friends. They met Harry. They all disapproved of me, and we didn't invite them again. When we had people over, they were my friends. Mac would sit at the table, listening to us but not really listening. He was always by himself.

Harry turned twelve. He was fat and pimply, oversexed, full of worries that girls didn't like him. He couldn't dance, so I tried to teach him the box step. He wouldn't learn. He didn't like dancing with his mother, he said. He always was unhappy, very fresh to us, stayed by himself. On Fridays after school, he either played stickball in the street with his friends—all in dungarees, high greasy pompadours, swaggers, Eisenhower jackets, and turtlenecks—or he did what he really preferred. Went to the library, came home with a pile of science fiction books, drank some milk and ate some cookies, went upstairs to lock himself in his room with the toy pistols he still kept. Reading fantasy trash in his room by himself.

For his twelfth birthday I planned a surprise party. Carefully learned his friends' last names so I could telephone them. Even the girls he liked. Made my plans with Mac. He said it would be nice. Nice. Because frankfurters made me sick, I wouldn't trust them. I shaped hamburger meat at night into long tubes. Mac brought home charcoal for the grill. On a Saturday in July I sent Harry on a complicated series of errands to Avenue J which would keep him away for an hour or more. He grumbled, of course. He was always grumbling, half in tears, fearful, mad. While he was gone, as arranged, his friends came. Three boys who looked like bums. And one was the son of a high school principal! And four girls, all looking the same. Fluffy hair, sleeveless flaring dresses that swished. One of them had breasts as large as mine. Mac couldn't look at her. He played his avuncular personality like a matinee part. Laughing with them, getting them to help him make the fire, setting out paper plates and plastic forks on the backyard table.

It was a bright hot day, the trellises which screened us from the neighbors were green and shining. Those wonderful high ancient trees rolled in the wind. The children had stacked their presents for Harry and were talking awkwardly among themselves. Mac pruned hedges and dug up weeds. He loved to do that. Sit on the soil in his army coveralls and push his fingers into the dirt. I had the cake in the icebox and was starting to cook. The beef smell and charcoal were mixed with the slow fire Mac had started in a garbage pail for weeds and old leaves. It got hotter and everything smoldered. The air in the backyard shimmered. I was happy, and happier when Harry drove home on his bike, carrying a high bag of groceries in one hand, braking and putting his foot down in imitation of motorcyclists he'd doubtless seen on TV.

His friends waved to him without speaking. Then one of the girls went up to him and kissed him on the cheek. They shouted Happy Birthday. They all shrieked laughter. I thought that he would weep. But Mac came up and broke the uneasiness, hugged Harry's shoulders, made one of his little unfelt speeches and then a joke. They laughed more easily, came to the table to watch him open his gifts. He followed very slowly. The girl who had kissed him was holding hands with one of the thugs. He rubbed her buttocks when they thought I wasn't looking. Then: aftershave lotion for my boy! A necktie thin as a string. A paperback book by Isaac Asimov from the girl who had kissed him. How could she know him so well and let a hoodlum rub her body like that? A child. Or a wise child. She made me angry and sad.

They ate in silence, the moment of emotion over—or the imitation they had resolved to create. It was over. Even during the cutting of the cake. They had sung "Happy Birthday" in cracked soft voices. Then they ate as if their embarrassment choked them. Harry looked at his book on the table. His face looked panicky. Mac was throwing weeds on his fire. I acted fast.

I went into the house for the ropes and showed them to the children. All cut into the right lengths. I told them some games would be fun. The boys smirked and raised their eyebrows. Pushing pimples up. The girls smiled at each other when they thought I wasn't looking. Harry sat at the table, red as if with sunburn. Finally he said "Ma. *Ma.*"

But I knew what to do. I called two of the boys to the far end of the garden while Harry shook his head and hung it. Mac stayed to watch, then went into the house. While I positioned the boys side by side and tied their adjacent legs together, Mac came back out and went to his fire. Then we were ready for the three-legged race. I called "Ready, get set, go!" and they hobbled through the garden to the birthday table and back toward Mac and me. One pulled the other forward, one dragged the other back. They didn't work together well, as if they pulled another body between them. They made rude comments and sweated, tripped. I said "Make the ropes *help* you. Some tough guys. You act like you're in a chain gang!" I untied them. Because I knew what preadolescents were like, fifty per cent sex maniacs, I made up two teams, a boy and a girl on each team. Better to let them feel each other's legs in your own backyard than at some party with stolen liquor.

I tied Harry's leg to the leg of the girl who had kissed him and given the science fiction book. For fun, I set them opposite the tallest boy whom I tied to the littlest girl. Mac went into the house and came back out. Harry was still bright red. The girl smiled a tolerant smile. She acted as if Harry were her brother. He didn't protest any more, and I thought to myself that the party would end up successful even if I had to run in a three-legged race my*self*. I said cheerfully "All right. Now you've got the idea. Let's see which is the best team!" I love parties, and I hadn't minded the hours and money spent to make this one work. *I* would make it work.

Mac went back into the house and returned as I started them off. Harry and his partner tripped right away and lay in the dirt as the other couple hopped and dragged each other from the fire to the birthday table and back. They won. And Harry lay with tears in his eyes. I could see them. The girl whispered to him and then sat up to untie them from each other. Mac went into the house and came out with Harry's basketball. He threw it over and said he thought they ought to walk over to Wingate Field and have a coed game. Two of the girls said ritual thanks as they left. The boys waved. Harry, red and moist-eyed, called "Thanks for the surprise birthday, Mom. Thanks, Dad." He said it as his back was turning, so the words went with him. They were gone.

Mac went into the house and returned to his fire. I looked at a piece of frankfurter-shaped hamburger meat which I'd left on the grill. It was charred black, it looked like a tiny piece of wood, smoking. The gifts were on the table, and dirty paper dishes, smeared plastic forks and spoons. Most of the fruit punch I'd made them was gone. I felt sick.

Mac came over from the fire and sat on the bench of the table. He said "I thought it was a pleasant party."

"Harry hated it" I said.

"No." He said it the way a father comforts a child.

"Stop it, Mac. It was a flop. That's that."

"Claire? Don't be so upset. He's growing up. It's awkward. That's all."

"That's all. That's all. I tried so *hard*. I *love* to make parties for him. I wanted this to be so *happy*."

He said "Well." Well.

I saw the afternoon again and hated it. All I felt was humiliation. All he'd wanted was to get away. I saw everything again. Then again. I turned my head. Mac always looked at me a certain way when I turned my head a certain way. We were doing that. Then I walked over to his fire in the garbage pail, which had more smoke than before. He called "Claire?"

I pulled one of the books from the fire. It was only a little blackened, only the fringes of some pages were burned. *Johnny Got His Gun* by Dalton Trumbo. I pulled another out and beat the embers from it: *Heroes I Have Known* by Max Eastman. More, then. Engels on the Manchester working class. *Progress and Poverty*. A study of Beatrice Webb. Some of the pamphlets I used to push under doors in the thirties on my way to work. The books of all that dreaming. All those years in the Village. High hopes and big talk and the sureness we were right.

I walked to the hedges and looked into them. All I saw were insects and droppings and leaves. The books made my hands hot. Mac said "It's the only way to be safe."

"So now we're safe."

"They're firing people who even went to meetings and never *joined*. We were *in* the Party, we were in the Nature Friends, we were in Youth Against Fascism. They're fronts, Claire. People are making lists. Would you like me to get fired?"

"*Fronts*. You sound like a German. Like a senator. Fronts. I never asked for a front. I *wanted* to be a Communist."

"But now you don't."

"Now things are different."

"Tell *them,* Claire."

"All right."

He sat and looked at me. I waited for him to argue, but he almost never argued. It reminded me of my father if he shouted. He knew that. We wouldn't fight. I stood at the hedges and looked at him and he sat at the birthday table and looked down the garden toward me. He held his hands out at his sides, like an old Jewish man in the neighborhood I'd lived in as a child. He smiled.

I said "All right." I went over and put the books back into the fire and then walked toward him and past him and up the steps to the dark kitchen. He came in with a tray of paper plates and cups, the leftover food, even the short lengths of rope. He always was meticulous. He was old. He put the ropes into the drawer where he kept his hammer and nails and pliers. Everything was taken care of now. Our family was safe.

TRAIL OF POSSIBLE BONES
1939

It mostly spoke of death. To him who traveled unequipped. To him who failed in the care of his body. To him without compass and map. To him who fell into exposure. To him who climbed alone.

    The thick carved sign which the Appalachian Mountain Club maintained said TURN BACK IF you were unequipped, or out of shape, or on your own, in little yellow-painted letters that were burned into the wood. But I was almost equipped, and my body wasn't all the way gone, and I carried a compass and maps. I'd come to climb alone, on the Appalachian Trail to the Mizpah Springs hut, a three-sided shelter that was three and a half miles up the steep difficult path where you start the climb along Mount Washington in New Hampshire, one of those eastern mountains where you can get surrounded in under a minute by solid fog, or ice storms, or wet thick snow, and wander with your dying body until you lie like a fetus on the rocks above timber line, killed by the wind.

I had come by train to Franconia, leaving without wanting to, not wanting to stay, and the gear I carried was barely enough, though I'd stopped for a little food. I left the hired car along the road and went to the mouth of the trail and read how, according to the mountain club, my chances were very poor. I rubbed oil of citronella along my arms and neck and face and under my socks and cuffs, and I tied a cloth as a sweatband around my forehead, and then I smoked a cigarette, felt the smoke stay down when I inhaled, and then went up a muddy incline, along a smooth round rock as wide as the wide trail, then over a bridge where the gnats clung low as smoke—the stream was low, it sounded tepid and foul—and then, a couple of dozen yards beyond the bridge, picked up the first marker, red circle in a white one, painted on a tree to the right of the trail. The trail closed in until the brush was always in my face and on my legs, and it stayed that way as I climbed.

It was hot, muggy in June, and the sweat ran into the citronella oil so that I was greased. After a little while I coughed up the smoke I'd inhaled, and that brought phlegm, the mucus of years in the city, the thickened air I had breathed for so long, and I stopped walking to cough more, then to drop to one knee—the pack almost pitched me backward—and I vomited on the Appalachian Trail as if I'd just come off ship onto land for the first time. I said "Earth sick" and quit talking, panted until I could breathe, then breathed in rhythms I counted off until I was settled enough to walk farther up.

The jungly forest around the trail was high enough so that I couldn't see anything but greenness and moisture. Ahead of me, the ground slanting up was wet, getting wetter, finally much of it mud. It slowed me, and I slipped, had to work too hard to walk, then had to rest too soon to have made any progress. I turned so that I faced downward and sat with the frame pack resting on a rock while I looked at where I'd been: it looked like where I was going. I snarled the air in and fought the tempo of my breathing down and then got up to walk before my legs got stiff or cramped. I went from trailmark to trailmark, and my thighs and buttock muscles continued to ache, as I knew they would for days. And after a while I stopped thinking about my legs, or the little streams that cut across the steep trail, or the bugs that came into my nostrils and settled at my eyes, or the consistent effort that breathing was, or the dryness of my sore throat, and I merely walked, conscious only of the walking, its stupid rhythm, looking only at the ground.

So that I had rested, walked on, drunk from a stream and filled the canteen, walked on farther and rested twice more, seeing only the muddy ground and the bull's-eye markers before I understood that I was talking to my wife, telling her again that I had to go away because an aunt had died whom she'd never known, whom I hadn't talked to for years, and that I couldn't explain why her death not only mattered, but came as a sneak attack: a kind of illness in the city which I fled.

I was resting again, at the foot of a rocky bank I'd have to three-point up to stay on the trail. The jungle was denser, as if it grew in close and thick for a final long surge before—the trees already were growing lower, shooting higher roots—the trail went up to timber line where the winds and cold made the trees grow stunted and low and then, above that, let only lichen and alpine flowers grow. My legs began to shake and I rubbed at my calves to slow the cramp, but it came in the right one, a long spasm; I pictured the soft muscle bending, twisting up and slowly collapsing into place, just as I, mimicking, bent and subsided. The air was thick with moisture, the sun was yellow and far away, distorted, as if the water in the air made a lens. I smoked another cigarette. I thought of Claire telling me not to smoke. I thought of Aunt Sarah, thought of not knowing her, thought of the last time I'd seen her: my father at her grave, holding my sister Ida as she wept on his chest, as I stood behind him and off to the side, as his little square face trembled and clamped, didn't break, shatter, into tears and the small-child's face of weeping adults. And I saw my face, as I fought not to weep for my father who fought not to weep. I thought of Claire, who hadn't been there, when I lit a second cigarette and put it out by rubbing it on a rock and burying it in mud. I said "Claire, leave me alone, please," and then, while

I imagined her reply, and thought of her comforting my mother, who hadn't come either, I started to climb the small cliff.

I wanted suddenly to be above it, because there was sky up there, open views, air that maybe moved instead of the water I'd been trying to breathe. I had to be above it because I was buried in jungle and mud, not breathing enough. I sucked what air I could and made myself dizzy, stopped trying to climb, counted large numbers to make myself steady again, failed, went up at the moisture-slickened rock as if a tide of mud were rising to catch me—fossil forever, open-mouthed, bug-eyed, always screaming underground—and found a hold for my left toe, wedged in, grabbed and lunged instead of fixing three points sure before I moved, and somehow caught. I caught again, and I went up where you shouldn't be unless you're roped and climbing with partners: for I had missed the trail completely, I found—had gone around it instead of with it, and what I'd thought was where I should be was a dozen yards to the left. I went up, though, missing all those bull's-eyes, and I lay on the rock shelf I'd aimed for, racked backward on my pack, panting, a hunchback pursued.

I went through my repertoire: being breathless; catching some of the breath back; drinking from the canteen; asking Claire to leave me be, stop asking questions, also stop watching in silence as if she disapproved and understood; rubbing my legs; touching the cigarette pack and sliding the hand away; taking one more drink; rolling onto my hands and knees; standing with a groan you only make alone.

There was no sky. It was the same jungle, the same muddy trail, just more rock, a diminution of trees, the brush edging close to the trail, the bull's-eyes off to the right which I had to catch up with. So I started again to walk, and soon I was breathing more or less as I should, looking at the markers and the mud, thinking of nothing, frying like potatoes in my own hot oil. And then the trail went down, not very steeply, but low enough for some different muscles to work in the legs, and when it started up again the legs felt better and my breathing somehow improved, and I felt very grateful for the fast-moving men and women who cleared these trails every year with brush axes not because of aunts named Sarah and distant graveyards but because they lived with their bodies in peace. I began to make time.

And then there was the drying of mud, then only soft earth and low grasses, the forest falling away beside the trail, the rocks increasing, the trees shrinking squat, the jungle feeling thinner, and then its disappearance and the sense of ridges crossed, the wind-cropped grasses and trees, the feel of rock sheets in the ground like massive long bones, and then, tilted into my vision, the topmost ridge and mountain meadow, Mount Clinton above me and to my left, the long purple sky and moundy Presidential Range floating under it like green and purple sea, and the mist coming in at dusk to the small three-sided shelter made of logs called Mizpah Springs hut which I had come to from Brooklyn in 1939, dragging my living and dead.

It was a wide mountain meadow, almost a bowl on the high ridge, and as evening came the short coarse vegetation, the boulders green with lichen, all began to glow with a buttery light. The earth grew rich and luminous, and then the sun disappeared into horizontal cloud banks and the clouds beat red like rows of cooking coals. I watched with my pack still on and smoked while my body cooled. My heavy boots made my feet feel like hooves. The ground seemed to shrink away from the sky, which was going out; the rock and earth and small bright flowers turned red and beat like a fire and then were merely pale shades of gray and yellow and green. The sky was unbridgeable distance.

I groaned as I stood, the cigarette made me dizzy. I thought of Claire telling me not to smoke, and I thought of not smoking; I thought of Claire telling me good-by, be careful, and I said "Claire, leave me alone, will you?"

The hut was made of logs—even the roof was made of logs split and bound and covered with tar paper someone had hauled—and I went past the ring of rocks used as seats around the fireplace made of heavy rock slabs with a metal grill across the top, stepped over the sill log, went past the cache of emergency rations and stacks of wood to see the floor of packed earth, the sleeping shelf of logs and fir boughs, sweet and dry and brown, and that was all. Except for the spring, a distance below the hut, surrounded by some kind of heather, this was Mizpah Springs, all that it offered. The mist was coming in now, wet on the face, cool in the nose and mouth. Early in the morning, when it would feel too cold to move out of my sleeping bag, the mist would be there, lying on me like silence, or snow. The pocket straps were made of leather and the laces of hide, the canvas of the shoulder straps felt thick as a grownup's thumbnail. With the pack off, my back felt filled with air, ballooning. I pushed myself down and rolled out the mummy sleeping bag on the bottom shelf, the rubber-lined canvas ditty bags, the canvas pouches of food and gear. Now they carry nylon packs of unavoidable red—mine was khaki—and nothing weighs very much. If I went back I would carry what they do, but now I don't go back.

I carried water up from the spring, dripping from the brown canvas water pail, and hung it on a nail in the hut's frame. I put my foil-wrapped bacon, the eggs in an aluminum case, the half pound of hamburger, all into the water, where nothing at night would get them and where they'd keep reasonably cool. Then I ate the two ham sandwiches I bought for lunch but hadn't been able to look at, drank water from the pail, put my things back into my pack, and went to bed.

That meant taking off my boots and heavy socks and cotton liner socks, wiggling into the bag and lying still. The mummy bag, filled with goose down and terribly warm, was shaped like its name: narrow at the bottom, wider at the top, going up past the shoulders and coming around at the head to curve almost over the face. It zipped from the crotch to the nose—you'd never freeze in one—and had a zipper pull inside which in the tangle of nighttime and dreaming was hard to find. And I couldn't stay inside it. Lying there, I always kept moving my hand to the pull, as if I'd be sealed away; every time I fell asleep to the jumping of my cramped muscles, I'd start, try to sit, be held, reach for the pull, push my face at the air. So when he came I was sitting with the bag zipped open, piled around my waist.

First I heard the rustle of bushes and the falling into silence of insects. Then I heard his footsteps on rock and dense earth: at night the ground banged like floorboards overhead. And then I heard his breathing, liquid and straining, high in his throat. I called "Hello the trail!"

"Hello the hut."

I said "Come on in."

He shouted back "I heard you, already. Hello." I shone my flashlight into the rock and small trees around the hut and he said "Okay. I can see, okay." I turned the light off and sat with my mummy bag around my waist like a woman in bed with the sheets around her naked hips, waiting. I put the flashlight away.

He stumbled into the sill and said "Fucking son of a bitch" and stepped over it—he was silhouetted against a glow of moon that made the rocks look icy—and he looked like an outline of troll: high floppy hat, pointy pack, bulges and bumps on a little frame. He said "Good evening" in his wheezy wet voice.

I said "Kind of late climbing."

"It is."

"Guess so" I said.

"Want some whisky?"

"Huh?"

"Oh Lord: are you in the Appalachian Mountain Club?"

"Why? Do I get a discount on Scotch?"

"Because those bastards are always shocked when you drink more than cider or smoke anything except a pipe full of good old woodsy Prince Albert or *do* anything up here besides get up too early and run your ass off into the steep trails."

"No," I said, "no. I smoke cigarettes."

"Wow" he said. "You're gamy. Would you like a drink?"

I said "No thanks" and lay back on my elbows and listened to him snuffle and snort, spit, yank things out that rustled, things that rang and clanked, swearing, kicking into wood and canvas. In that silence, he was a stone dropped into placid water, making a roar of ripples and foam.

Pretty soon he was quiet, still standing near the food cache, and then he said "You have any objections if I don't take the upper?"

I didn't know if he was talking to me, I lay still.

"If I sleep down here, where you are—a very discreet distance away. Everyone up here is always very discreet."

"Oh. Me?"

"No, my friend, the naked woman I lead around by the chain I keep on her neck."

"Yes, well no. Or yes. Whatever you want, go ahead."

He threw more gear and swore at it, crunched across the boughs, stood outside—tripping on the sill as he went—and peed loudly on rocks. He came back in and dropped with groans in several voices onto the sweet branches and was still again. I heard the bottle being capped and I fell asleep at once, as if I'd waited for him to bring me rest.

His screaming woke me—very high and thin, a new voice, that of an ancient child. He cried "Noo. Noo. Oh noo. Oh dear. Noo."

My fear made me hold my breath, my choking made me think. I pulled the mummy bag up from around my waist and held it to me from the inside with my fists bunching the cloth. I closed my eyes in the dark and said "Mister! Hey!"

"Oh dear" he said. "Noo. Oh dear."

I shrieked "Wake up! Wake up! Wake up! Wake up!"

He shifted, jumped, lay still. The breathing was even and liquid, unwell. He very calmly said "No reason to shout, friend. I'm sorry my snoring woke you."

And I sat there while he burred and purred, snoring now of course, and I wouldn't close my eyes. Like Ida, my younger sister, who when she was put to bed by my parents said so often "I'll sleep, but I won't close my eyes. I hate to close my eyes. I never close my eyes when I sleep." I fell asleep sitting, fell down, wakened, rolled tighter into the mummy—still not zipping it shut—and slept with my fists clenched into the cloth on my chest.

I didn't dream of Aunt Sarah. I heard my parents talking with foolish delight, and I felt ten, and then I smelled wood smoke, the chilly charged wetness of in-coming mists, and opened my eyes, thinking before I sat that the other smells—cheap fatty bacon, scorched eggs—were from what I had stored for my breakfast the night before.

He didn't mention my food and neither did I. He handed me a paper plate full of greasy eggs and burnt bacon, and I sat on a rock, shivering in my wool shirt, looking from under Mount Clinton's brow over the mist that steamed away, the sun glinting off rocks and even ice in the blue distance, the cry of a gliding high bird that seemed to pull at everything and fix the world in place.

He ate with dainty motions, but with his fingers and a muffin—he'd provided it—which he sometimes dipped in the panful of grease, and which he sometimes used as a shovel for wet eggs. He poured us coffee—his treat too—and looked away at the mountains as I did, tasting them with his food. He had eyebrows so thick the rich brown hair curled, and his nose was wide and shining with grease. His yellow-gray hair fell onto his face and neck as if he wore an animal on his head, and his ears and nostrils sprouted. He was a trim taut man who breathed very badly and looked very angry about his old age. His eyes were wet, and they gave him away: he looked as if he'd just cried.

We put the paper plates in the low fire and sat at it, drinking coffee while the day kept coming in. We smoked my cigarettes for a while, and then he said "You look at me like I've got *schwarze* blood. You think I'm a Negro person?"

I said "No, I thought you were very tan. I never saw a black man with blue eyes like yours."

"I've got lovely eyes" he said. He took another cigarette from the pack on the rock beside me. "So you've got me pegged: I'm not a *schwarze*."

"You're a tan Jew."

"They're the same?"

"Nothing's the same. My wife says that. No: it's that you know how to say things in Yiddish."

"You forget your heritage?"

"No."

"You deny it?"

"Often. Yes. Or I don't bother to remember it."

"It's hard to forget."

"Everything's hard to forget" I said.

"Oh. *Very* meaningful."

I said "You don't see too many Jewish mountaineers nowadays. Or even regular hikers, like us."

"I'm not regular" he said, shifting his bottom and raising his brows. "Never have been."

"I mean most of the people you meet on the A.M.C. trails are more or less . . . regular Americans."

He said "No one's less, everyone's more. You're right. The joint usually buzzes with them."

I didn't answer. I lit another cigarette—I thought of Claire shaking her head—and I looked at his tanned brown hairy fingers, short on wide hands; his bare legs under brown shorts were heavily veined, like Claire's, and the calf muscles bulged as they should for the sort of walking we'd done.

"So, what are your thoughts about the world?" he said, lighting one of my cigarettes.

"I try not to think about it too much. No, that was a wisecrack."

"No kidding?"

"It's very frightening."

He nodded, then said "You have a father alive?"

I nodded back.

"Around my age?"

"He's not so active any more. He was a carpenter, a very good carpenter. In Russia, then in Poland. Then he came here and worked, he broke his heart carrying bricks up ladders for Americans like an apprentice."

"The usual story."

"It wasn't, for him."

"No. But he's not in Europe at least. He might have died by now. He will, of course, but not killed by a gangster."

"He isn't well."

"Who's well? In Europe the *healthy* ones are dying. They'll move on Poland. Schmuck Hitler says he wants peace, but he really wants Poland. And Czechoslovakia. And France. I don't know about England. But if Poland goes—"

"England?"

"Sure. Then us. Maybe."

"I don't know . . ."

"And by then or a little bit later, all the Jews will be dead."

"No."

"Listen, mister, I don't want to be right for once."

"So we'll have to fight."

"For Jews? Nah. For war debts from twenty years ago. And for rubber and oil. For *that* we'll maybe fight. You'll fight. Would you?"

"I'm going to have a baby. My wife's going to have a baby."

"Ah."

"And anyway we're neutral now."

"Not me, pal. I'm walking around while I can by myself. But I'm thinking, understand?"

"About having a war?"

"About the gangsters and the Jews and the dead babies. The world will fill up with the babies they kill until we have to build roads over their bodies. I hope to God I live a long time. But I also hope I am struck dead before I see what happens. A little bit, I hope that. I would also like to get laid by a fat juicy whore one more time. So. I'll leave you with this mess to clean"—his accent was more Boston than New York, and sharp with ironies, and not displeased with itself—"because you're the young apprentice. What're you, thirty yet? Thirty-two? And I'm on my way. I'm going the slow way up Washington, and if anybody asks you in a week if you saw Samuel Dreff, the retired accountant, you can claim the pleasure and tell them where to look."

On the trails you always tried to tell someone your destination in case you were lost or hurt and they had to come for you. I said "Mac Miller. And I'm going the same way later. I hope you enjoyed the bacon, which isn't kosher, as our heritage makes clear, and I'll look for your bones. I hope I don't find them."

"That's classy" he said. "Very classy."

I said "What'd you mean: I'm the young apprentice?"

He said "Apprentices never ask questions unless the *maître* —there are French Jews too, though not so many, and dwindling—unless he asks to hear them. Be happy with what you've got. Good luck. Any kind of luck these days is something." His face drooped from its raised-brow poise for parries, and he went into the shelter—tripping over the sill, shouting "Son of a bitching bastard!"—where he packed his load on an old plywood packboard and waved and went around the back of the hut to climb along the side of Mount Clinton and onto Washington's shadowed side.

I poured the coffee out of the small aluminum saucepan (mine) he'd boiled it in and heated water for dishwashing. As I poured the bacon fat into the little fire and watched it roar up, I remembered how many times—especially the last few weeks, when Claire's leg veins were very bad—I would stand at the sink in our apartment, the Ocean Avenue traffic subsiding, its noise a trailing out around eight and the promise of silence actively in my mind like a reward for punishment borne. The little kitchen would diminish about me, the talk on the radio would become softened inarticulate sound. In a kind of blackness I would see only my furred arms sloping from the rolled Oxford shirt sleeves into the suds, and my hands would not be there. I would stand and stare at the absence of my hands and see how I might never grasp or seize, and never be required to. Then the noise would come back and I'd pull my dripping arms, slimy, up and into the air.

At Mizpah Springs I washed the dishes of Miller and Dreff. I'd kept the frying pan out, and I laid my half-brown hamburger meat in patties in the pan and fried a lunch to eat on the trail. I made hamburger sandwiches on which by afternoon the bacon fat would congeal so that my lunch would taste like breakfast. Communion. Then, after I'd wrapped the sandwiches in the foil from the bacon, I put the dishes half wet back into the canvas sack, poured dishwater on the fire, kept two pieces of bread for an emergency—along with a bar of chocolate, a compass, an army-surplus whistle I'd use to call for help, a rubber sheet I could wrap around me and my sleeping bag if I bivouacked, lost, in a storm—and I left the bread behind with the cache of provisions to which everyone added before he left. Dreff would have said "Communion for the regular American mountaineers." I rolled my sleeping bag into the knapsack, tied the water bucket on the back to dry, filled my canteen, read the map. Then I collected some wood from a couple of hundred yards back on the trail and left it on the stockpile for travelers. Most people left more behind, and covered more miles in a day with their earlier starts. But then, of course, they weren't running away, I guessed, crippling their progress with fear. I guessed.

My pack was badly loaded, and it wobbled on the rock I was backing up to in order to get it on. I said "What fear?" the way Claire had said it, round with attempt Number Three —surprise!—and sitting at night in our living room, Flatbush a sleepy suburb, her husband dressed in boots and heavy pants, carrying a knapsack she hadn't seen since we'd moved to the apartment.

I sat across from her and drank black coffee to keep me awake on the trip at night. I smoked too many cigarettes. I said "I told you about my aunt. Something bad happened to me when it happened. I *told* you that."

Claire said "Don't get yourself angry. I think you're trying to get mad so you won't feel bad when you leave, Mac. You'll feel bad anyway, no matter what you do."

"I'm not."

"*I'm* not."

"I know. But why?"

She said "Why I'm not mad? Because you're taking off in the middle of June and I have to call the college and tell them there's a death in the family?"

"Well there is!"

"Aunt Sarah. Beloved Aunt Sarah."

"I thought you weren't mad."

"Don't pick a fight with me, Mac. You want to go? Go."

I said "And you're happy I'm leaving you alone and I'm running away to New Hampshire?"

"If you want to do something so badly—"

"And you're happy with my explanation?"

"No. Aunt Sarah? No. I don't understand how she's the explanation."

"Claire, so how is everything all right?"

She sat with her small bare feet flat on the carpet, a chubby woman poised. Her face was white and she held the muscles still. She said "You're coming back."

"Of *course* I'm coming back. Dammit. Don't be so dramatic."

"No," she said, "I won't be dramatic." She waited, and I did, and then she said "Mac, you're stupid. I don't hate you. I'm not mad. But for a smart man, you're stupid."

I stood at Mizpah Springs with my pack hanging heavy to one side, looking up the trail that curved behind the shelter and along the side of Clinton, then over to the lower slopes of Mount Washington, where there was nothing to see but rocks and clouds and where the trailmarkers were painted on pyramid cairns. That was where the danger was, where the moisture froze in the cracks of rocks and split them, where the mists and wind appeared and vanished, where people sometimes vanished.

I thought of Aunt Sarah, my father's sister, and inside my mouth Samuel Dreff's voice said "Bullshit, my friend." My tongue stuttered back on the words and I said them again, to pregnant Claire who was thirty-two, like me, and who was going to enjoy her final weeks of pregnancy at the height of the city's heat. I said, in my own voice this time, what she'd never told me, and what I knew, and what I hadn't told her: "I don't want a baby. I don't want a baby coming again." I looked at myself as I hid in the soapsuds, huddled from my neighborhood's sounds. I looked at her in bed, sweating, a mound of dread, an egg of anxieties and sleeplessness and the guarantee we'd never be alone until we were old. We were old. Claire told me, and Dreff, in his voice, "For a smart man, you're stupid."

We were old. We were catching up with my father and Aunt Sarah, Samuel Dreff. We were dying. So I had to return, sliding down the mud and into the dense brush away from this sky, and I had to travel home to Claire and tell her, and try to have our child, and what of *us* we could keep protected.

I stuck my hands out behind me and jumped in place until the pack shook down on my arms and fell to the ground. I walked away from it and looked at nothing, walked away, then came back and walked past it in the other direction. I breathed so deeply, I thought I would faint. I sat down with my legs before me on the stones and scrub. I stood, still breathing loudly, listening to my breath, then I walked in the other direction and then came back and lugged the knapsack to a high rock, propped it, backed into its straps and, with my arms folded in front of me, I bent over low and walked up the trail without resting until I was above the hut, at the verge of actual timber line, where there was no shade and the mica threw back jewelry dazzles of raw sun. I stood, facing downhill, resting my legs, thinking of the bones I had promised to watch for.

So there were instructions to follow. Go the rest of the way through the boulder fields. Make sure Dreff had survived. Climb back down a faster route and hitchhike around to the car and return it and get on a train to New York. First be a Jew of sorts, follow the trail of possible bones. Then be a sort of a husband, go home.

TAX
1953

Harry said "Moon? See moon? Mommy. Daddy. Moon go *way*."

    Once my parents tried to live on a farm outside of New York and stay healthy, but the project failed, and so did their bodies, and I grew up in the streets of neighborhoods in Manhattan and the Bronx. We drove back there one week end, while I still had my license and nerve enough for the roads, when Harry was very small and Route 17, near Tuxedo Junction, retained a sense of difference from New York. There were hills and long green stretches, and exhaust didn't hang on the meadows. But that's all it felt like: a different, greener place. We were driving home after the picnic, I was squinting into the six o'clock sun and seeing silhouettes too late, though they kept missing us and going past to vanish on the road that looked like mercury behind us in the low final light. The wheel shook my hands and the car drifted sideways when I let it, fast cars behind us kept coming, as bright as if they burned; they overtook us, vanished just in front of us, disappearing into the light, consumed. I said "I didn't know you could see the moon this early, Harry."

Claire, with him on her lap, said "Yes. Just a little. It's to the left."

I said "I can't look left, they're all trying to kill us in a straight line."

She said "I'll tell you what it looks like, if we survive."

Harry said "Moon go *way*."

"No," Claire said, "the clouds are covering the moon."

"Cowds."

"Clouds" she said. "That's right."

"Cowds coughing—"

"Clouds" she said. "That's right, honey."

Harry sat and then he said "Look honey! Look honey! See? Moon coughing cowds!"

It's been a long time, now, and now I rise from the Brooklyn subways on a night in summer to tell my father good-by. Claire is in the hot close corridor of the home, her handbag on the marbled asphalt tile. The bag's soft leather has collapsed at her feet, and the handbag looks like something dead. Her chubby legs are locked to keep her up and the kneecaps bulge in knots, her thighs and stomach jut, her legs are spread. Above the waist she is weakness, the chest fallen back and the shoulders slumping caved beneath a swaying neck. She has no face.

When I'm still at the stairway door she whispers "Not dead. He isn't dead."

I whisper back, past closed wooden doors in a light green cinder-block wall, "You mean it could be so soon? It's really—"

"Really" she says. "In a nursing home."

I come up to her and put my briefcase on the floor beside her bag and we look at his door. I say "I know it's a nursing home."

She says "Then we all know it's a nursing home."

I say "Where's the doctor?"

"In the room."

"Very big deal" she says. "You sound like a frustrated tycoon. It's a pretty decent job."

Harry says "Do we go in and talk to Grandpa, Daddy?"

"What?"

"What do we *do?*" he says.

I put a hand out for him and miss, and the hand is back at my side by the time he shuffles closer to be touched. I say "I'm not sure."

"We don't know" Claire says.

I say "We don't know."

Harry says "God."

"You're fine" Claire says. "You're doing it very well."

"Big deal" he says. Then he says "I don't know what I'm doing."

Claire says "Did you have supper?"

Harry says "No."

"I left supper for you, honey."

"Mom—"

"We'll get something later" I say.

Harry says "Later? Well what if—what if we—"

I tell him something and he says something back, but the door is opening in and the doctor in fawn-colored golf slacks and a gold knitted shirt that shows loose breasts comes out with no face. I say something urgent to Claire, and I don't move. Claire says something and Harry comes close to my side and I say Harry, I'm all right, don't be scared, but I hear no words. He is standing beside me, fat and invisible, and Claire has entered the room. I watch her pushing the door shut and her fingers disappear from the edge. I look at the door.

I look at the door and it opens. Harry holds my wrist as if I'm a wobbly child. He is shaking very quickly, and the heat of his hand is honey that thickly settles through the cloth of what I wear. Claire says "Go in and kiss him."

Harry says "Me?"

I say "No" and no one moves. I say "No" and feel Harry's fingers lift away. "Me" I say. "Right?"

Claire says "You."

I walk to the door and she says "The doctor will stay there with you. All right?"

"Claire," I say, "I love you. I love you very much. I love you."

Her hands are hot on my face, and wet, heavy. They move away and come back. She says "Of course. Of *course*."

I say "Do you forgive me?"

"For what?"

"I don't know what."

Her hands are on my face and she says "Go inside and kiss him. Then come out." She says "Can you?"

I say "Yes." Then I say "I don't know."

"Well go ahead" she says. "Go in."

I say "Good-by" and walk inside and feel the door make a breeze as it closes. I see the metal bed and light green walls and the shape, the golden colors of the heavy-breasted doctor, round and ripe in a corner chair like summer fruit.

I hear the lungs snore and snuffle, separate under flat white sheets from the skull above; there is nothing holding the parts together now. I see eyelids of bone and the beaked nose sharp and blue. There isn't flesh, there is clear hard bony glaze, like fingernail, which has folded into sockets and melted onto the forehead and jaw. There is the white flatness of sheets, too low to be covering a chest and groin and legs. There is the head, sinking in upon itself, like clearest tallow melting down. There is the snore of lungs. There is the hush and sigh of the door as Harry comes clumsily in. I whisper "*Two* goddam buckets of fruit" and rest my chin on my chest and close my eyes.

Harry whispers "Daddy. Daddy."

"This is my father" I say.

Harry says "Daddy. Come on out."

I say "I'm not crying."

"Please come out" he says.

On the metal lampstand next to the bed are his fat gold watch made in Europe before I was born, and the wire-rimmed glasses that look too small for a face, and the wallet made of brown textured plastic, holes punched through the edges by Harry and laced by him with brown plastic thongs. I open the wallet and look inside and put it down. I take my leather billfold from my suit and I pull my paper money out, hard and dry and smooth. I open his wallet again and put the money in and fold the wallet closed again and set it down. I say "Good-by, Daddy" and hear his lungs snore, hear Harry snorting too as he pulls on my wrist with his heavy wet hand. I say "Good-by, Daddy" to the skull and the flat white sheets and then go out.

Harry is behind me, saying "Bye, Grandpa" and then we are both in the corridor, waiting for Claire who has gone back inside.

When she is out again we walk to the stairs. Claire holds her leather bag and Harry, with my brief case, walks down behind us, sounding loose and moist and alone. Claire says "They'll call us, Harry, don't worry. Okay?"

"Okay, Mom."

"We'll wait outside where it's cooler and they'll call us."

"Mom, *okay*. Okay. I get the message, okay? Okay, already?"

I say "Take it easy, Harry."

He says "Thanks a lot. Big deal take it easy."

"Probably I always took it easy" I say.

She says "You mean about your father?"

"My father."

"We did what we could."

"Oh, you don't *mean* it, Claire."

"No" she says. "I love you. But no."

"No."

And we stand outside the low stone building at night in our city. Traffic goes both ways before us, bright as it comes toward us, red and bright as it goes away, then red and duller and gone from sight. Street lights on brown steel stanchions over the black avenue droop like dangerous fruit in a foreign place. Claire holds my chin and turns my head toward her and I see no face. I see houses and closed stores, glass and brick and metal and the shape of her as she sees the shape of me at night, in summer, as we wait.

She moves away and then I hear her giggle as she says "You see the moon?"

I say "No."

"Did you look for it?"

"No."

"Well you wouldn't see it if you did" she says.

"Claire, I don't understand a goddam thing you're saying!"

She says "Because the moon is covering the clouds."

"Oh."

"The *moon* is covering the *clouds*. Don't you remember?"

I say "Oh. Yes. Yes. Harry?" I reach for him but he's at the edge of the street, away from us, my brief case at his feet on its side. He dances slowly, around, around, a chubby spotted discus-hurler winding up beneath the sallow lights of the nursing home windows. He stops, his own brown plastic wallet in his hand, and weaves as if he's dizzy. I call "Harry!"

He says nothing. I can't see his face. He turns again to the street and spins again, sails his wallet into the moonless night above the traffic coming on, above the pallid light that leans from windows in the nursing home, up to the night as if its blackness were a sea his wallet dropped to from enormous heights. And the wallet is gone, Harry has staggered to sit on the curb beside the traffic, next to my overturned brief case, palms on his knees and his elbows cocked, his head leaning down as if he looks from his heights to the sea.

I say "Harry."

"Let him sit" Claire says.

I say "Harry?"

She says "Honey—let him."

"Let him *what?*"

She says "Just let."

I say "Harry. Hey." I say "Listen. Harry? Look: thank you. I'm sorry for it. Thank you very much. Harry?"

A car goes by to light him up and, passing, shuts him

off. Another goes past and he is lit and then extinguished. He says nothing.
　　Claire says "The two of you."
　　"Harry?" I say. "Thank you very much."
　　And he says "Yeah. Big deal."
　　I say "Big deal yourself."

TWENTY-ONE THOUSAND AND CHANGE
1959

The vows they never spoke aloud as they drove from New York, choking over the West Side Highway, rolling sideways off rounded cobblestones into dangerous lanes, not watching the Hudson River gather itself always and move: no reference to the fact that four years of college had cost them twenty-one thousand dollars and change. No description of the length—three hundred miles—and duration—five hours—of their trip from Riverside Drive to the center of the state. No shocked description of him ("You're so *skinny!*")—long hair, short hair, dueling scars, nose tattoo, anything: just say hello and don't force the kiss. Nothing to be said about phone calls home or letters in the course of the school year past. Take both days of graduation week end and then see, we'll see, we'll see. And not until the end of it all—The President's Father & Son Golf Tournament, The Merrymakers' Concert of College Songs, The President's Reception (4–5 p.m.), The Dean of Students' Seminar ("What is a Liberal Arts Education in 1959?"), The Dean of Faculty's Seminar ("Educational Innovation in a Multidisciplinary Proficiency Degree Program"), the play on Saturday night (*As You Like It*), the brunch on Sunday, the Commencement (heat and dampness, the faculty marching to perky organ music, flashguns and infants exploding, the Ambassador to the Netherlands, the retired singer and the eager college theologian and the awed bored industrialist there to receive their honorary degrees, the gymnasium emptied out later on and littered, the stink of departure, the feeling of being alone at a terminal after the trains are gone)—and not until the end asking whether two will go home or three. To get through the bedrooms and bathrooms and living rooms and classrooms, dining rooms, ceremonial rooms, inner rooms and outer rooms, and then to see: we'll see.

Their room at The College Inn was furnished in Colonial Veneer Moderne: wallpaper showing American dolls shooting muskets at Hessian dolls from behind a neatly trimmed hedge, over and over, around the room; the bed and bureau with rounded bulges of printed-on pine knots which were heat- and liquid- and sun- and stain-resistant; the bedspread showing Indians in blackface dumping tea that looked like hay into a septic harbor; the tangs of mildew in the bathroom; the bath mat—heavy paper—on which were printed Washington's head and Jefferson's house; the sound in their corner room from the Pub Corner one floor below—a Jewish girl from Mount Holyoke singing up from the juke box, to an amplified bass, that the Rock Island Line was a mighty good road.

Claire opened the suitcase and took out some underwear and sport clothes and crushed them onto the shining bureau. She left them there and sat on the bed very stiffly. When Mac came out of the bathroom, pale in the brightness of the overhead lamp, he brushed his fingers down the drawn blind and walked to the bureau, put the bunched clothes back into the suitcase and flipped its top down: a mouth too full to close. Then he looked at the suitcase, nodded, looked back to her. He said "Well he got us a room."

She said "He wants us here. Really. Don't you think?"

"I would say I have absolutely no idea. I don't know—yes. I guess. I guess if he has to be here he figures we should be here. No: probably he's figuring he owes us this and what the hell, we might as well get a collegiate hotel room and really enjoy what he's going to hate for forty-eight hours. I don't know. Do you think he thinks we *like* this?"

Claire sat stiffly and moved her hand on the bedspread. "I keep wanting to vomit" she said. Then she said "Did you see what you did with the suitcase?"

"Again."

"Always. I'm always taking things out and you always put them back."

"And we know what that means" he said.

She said "Well this time I don't blame you. If there were enough room I'd get in there too."

"I'll meet you inside."

"I'll see you in a couple of days" she said.

"Take care." They stayed where they were.

Then: downstairs in the Pub Corner, the large high room of wood all sweatened with Saturday's mugginess and the heat-moistures of nearly a hundred people drinking and talking and moving in place. The room was so crowded, an extra wooden table carved with initials and fraternity names had been set on the flagstone platform in front of the great stone fireplace over which a mural of the thirties showed bearded men at work and broad-shouldered women watching them under a pure sky. Mac and Claire and Harry sat at the extra table and were elevated by the flagstone terrace. The juke box was loud. And each looked over a shoulder—raised, slightly, and stiffened—at the room, the beer taps, the parents and children drinking manhattans and martinis together as if it were a business day in some sizable city and everyone was everyone else's colleague after work.

Claire said "You look fine in your sport coat, Harry."

Harry said "You look nice too, Mom." He looked at Mac and said "How about you, Dad? Uh-huh. You qualify too, Dad: you look good." He put his palms on the sticky table and smiled his yellow teeth and everyone smiled and drank a sip of drink, looking over shoulders at the room.

Harry was very tall and suddenly thin, his long acned face haloed by a fan of curly uncombed hair. He wore a tee shirt underneath his ocher corduroy jacket, and it showed the pulse in his pale throat going quickly as he smiled and moved his droopy fingers on the table and held his body still. He was larger than his parents, but they all looked alike. They all sat still.

Mac said "Where'd you get the sport coat, Harry? I don't think I remember it."

Harry said "Boy, it's really a hit, huh? Yeah, well I bought it from a guy. He needed the money and I needed a jacket so we did a deal."

Claire said "I didn't know you needed clothes. We could have—"

"Mom, it's all done."

Mac said "Right. Hey. So what's the schedule, Harry? What are the plans. Do we go see Shakespeare tonight or what?"

Claire said "I have the feeling Harry has something else to do."

He smiled his tartared teeth and said "As a matter of fact, Stacy, this friend of mine, and some other friends of mine, we're having a party tonight. At the house? So I thought you wouldn't mind it if I skipped the play."

"No sir" Mac said. "Going to see Shakespeare isn't—"

"I think *we're* supposed to go to the play" Claire said.

"You're not *supposed* to go, Mom."

Mac said "Is it kind of a private party?"

Claire said "And we're supposed to meet you tomorrow? For the convocation?"

"Would you like to do that?" Harry said.

Claire said "No."

Mac said "Why not? Sure."

She said "All right."

"No, Mom, why'd you say no?"

"Because I'm getting emotional or something equally reprehensible. No, it's fine."

Mac said "Fine, Harry. You meet us here for breakfast—hey, how about dinner? Let's go out someplace and get a steak. Are you hungry?"

Harry said "Look, I don't want to be crappy about this. It's just these things came up—"

Mac said "No, this gives us some time to walk around the campus. I haven't been here for a while. I've barely seen Mother all week. No, it's fine, we'll go get dinner and walk around and get some sleep and we'll meet you."

Claire said "Fine."

"Mom?"

She said "No, Harry, don't change your important plans. We'll see you later on if that's what you'd like."

"I don't want to be crappy. I mean, you spend a fortune sending your kid to school, you have a right to sit down to dinner with him and have some fun with him—the whole graduation business, you know. It's just this situation came up and I think I have to be there. Otherwise I wouldn't be so draggy, okay?" His throat jumped.

Mac said "We'll meet you later on, Harry."

Claire said "Have a good time, Harry. You look very nice in your sport coat."

Mac and Claire stood at a low wide spreading tree in the small quadrangle, at the other end from the white lighted Methodist chapel. The grass was wet, the night sky low and seamless, the lights on the dormitories steaming with bugs. Claire said "How old do you feel?"

Mac said "Ageless."

She gestured at Harry's invisible life, and at Harry who wasn't there. "This makes you feel *young?*"

"It makes me feel dead."

They walked beside the artificial lake which was fringed by shade trees and circled by a lawn so kempt it looked artificial as well. Two great swans circled, and geese hissed like snakes beneath evergreens. Mac said "They rent those swans. A dog killed the swans one spring when I went here and some vice president made a notch on his pistol grip by deciding it was cheaper to rent than to buy. So every year they rent a pair of swans. Which might contain a moral lesson for us all."

She sat on an unpainted bench that was, according to its plaque, memorial to a dead dean. "So what's the lesson?" she said. "Rent a son?"

"Nah. We rent a dog to kill him." He was snorting purposefully, as if he laughed, but Claire was walking away. "Oh, come on. It was only a tasteless joke. I'm often tasteless."

When he was beside her again, when they were walking around the lake, she said "I don't want him dead even in a joke. The point is that it's getting to *feel* as if he were dead. Or very ill with a special disease we don't know about. That makes him so skinny, and holy, and he can't speak the same language any more. Oh this is so terrible! It's so typical, it's what they make fun of in *New Yorker* cartoons, our own *New York Sunday Times* feature on alienated youth. I really hate myself in this." They sat at another memorial bench, and truck traffic ran south, the chapel bell—far away, up a high hill past the old library, the new library, the dormitory for freshmen —struck a blurred number of hours. Moonlight washed like bright water above them on the bottoms of the low clouds, but the moon itself was invisible. She said "I'm not very happy being who I am now."

"So much of you was involved in being his mother. Which is right, I'm not talking about anything at all within any possible definition that was wrong. And I'm not saying I think you should have done anything differently. I'm—"

"It sounds as if you're building up to something terrible."

"No, I'm only saying that now you have to learn to *not* be his mother. Or to be his mother differently."

"How?"

"Christ, I don't know. I feel the same as you."

"How can I not be his mother, Mac?"

"Be his mother from very far away."

"How are you doing at being his father from very far away?"

"I'm not. Right now, some twenty-five year old childless free-lance hack is writing an article for the *Saturday Review* on understanding our post-Sputnik estranged youth and how puzzled their parents are, and I'm in the middle of everything he says."

"So what do we do?"

"Change, I guess."

"How?"

He put his hand on her thigh and took it off. He put his arm around her back along the top rung of the bench and cupped her breast, kneaded it. He leaned in and said "Let's go back to Ye Olde Inn and fuck."

She sat stiffly and said "Since when do you use that kind of language?"

He took his hand away and put it on his thigh. "I suppose I never do."

She said "You didn't really want to, did you?"

"No."

She breathed out slowly. She said "That's good."

Away from the lake, beneath the hill on which the new split-level library was—it looked like a huge Long Island ranch house made of the college's granite traditional stone—they walked beside the small river that ran across the lower lawns of the campus. There were giant old willows and multistemmed pines, strokes of light and gashes of dark shadow on the darkened grass. The low sky glowed. Claire said "You didn't want to go back to the Inn, did you?"

"No."

"We're a little past making love for proof that we can be alone."

"Do we need proof?"

"We're past making love" she said. "I don't want to talk about it."

They stopped at another memorial bench but didn't sit. He put his foot on it and leaned an arm across his knee and Claire folded her hands. She said "Who do you think Stacy is?"

"He told us—one of his friends."

"He always called them his friends. Boys, girls, it didn't matter, they were always 'this friend of mine.' As if he didn't even want to give away their names. I'm surprised he told us about Stacy."

Mac nodded. "I wonder what Stacy is" he said.

"I don't understand. Oh. No, Mac, really."

"Well. Everything's changing. Everywhere in the world. Why not here?"

"You're saying Stacy's a boy and Harry's a homosexual?"

"No. I'm not. No. But I wondered. Isn't that incredible? I watched his testicles descend, and now—you know, it slimed its way across my mind. What you were saying."

"No."

"No, I don't think so. The worst is that I thought it."

"Could you get used to that?"

"Could you?"

"It's not called a sickness any more by many intelligent people."

"By queers it's not."

"It's a preference, Mac. A choice, is how we're supposed to think of it. But I don't think Harry's like that."

"Like what?"

"Queer."

"I wanted to hear what it sounded like when you said it. It sounds very strange. In fact, it sounds queer."

"Will you stop making those terrible *jokes?*"

"Yes." He put his foot down and looked toward the river. He said "But do you think you could get used to it?"

"No."

"No, neither could I. I hope we don't have to get used to anything else."

Claire said "What are we used to *now?*"

"I guess what we were used to before we had him—you and me."

"But then we had him for all those years before college. We're changed, aren't we? You know we are. I don't think we can be alone the same way any more. Whatever that was. I don't think we can be the same. He changed us."

He walked to the river and Claire came behind as he leaned down and pulled at a black branch that stuck up from the slow water at its silty edge. He pulled harder, then planted his feet and groaned, cut the sound off, pulled, then fell back, caught his balance, held to the branch, pulled, and the long slimy branch came up, then its wider end, then the long log it was part of—the river sucked and plopped—and he hauled it all to the grass and panted as he watched the water bleed off. He panted as she watched him and then he choked on a breath and rubbed his chest. He turned around, said "What I'm wondering." Then he said "Is did *we* change *him.*"

She put a hand on his moving chest and held the palm there gently while he wiped his hands on his sport coat. She said "That's the kind of remark you'd tell me to stop making. You'd say stop begging for guilt."

He said "Well there's guilt going begging tonight."

Claire said "Let's go back and make love."

He rubbed his wet hand or her white arm, rubbed and smiled, looked past her, said "Let's go back and have a drink."

She said "And talk about something else."

He rubbed with his seersucker sleeve at the powder of algae and bark he had left on her arm. He said "Something. We'll see. Let's go back and we'll see."

"And go to bed together?"

"If that's what we do."

"It's all right, Mac. Let's just lie down and talk."

They sat again in the Pub Corner, this time at a table in the center of the room, and they drank Canadian ale, moved their Schaefer napkins, and watched each other watch the students come in—most of them alone, some with brittle parents who didn't move to a table or go to the bar until their children made the suggestion, but most of them alone: in jeans and knitted shirts, their bodies usually slim, the breasts of the girls very evident under thin cloth and ignored by the girls as if they expected everyone else to ignore their bodies too, or as if beyond that pretense they knew very well that no one could help but look.

After a while Claire said "I can't stand this."

"Which this?"

"I'm not hearing what I say to you. You're not hearing what you say to me. I don't know what you're saying, and you don't know what I'm saying, and we're foolish, we're uncomfortable. We're not saying anything worth saying and we're not finding anything to say about Harry and we are generally disgusting me."

"We're doing what we can. We just have to get through this. And tomorrow. And then we can go home. And I don't think he's coming with us."

"I know."

"And we just have to *do* it, get through to the other side of it and go home."

"And then?"

"Get through to the other side of that."

"And then?"

"Ditto."

"For the rest of our lives? Ditto?"

"No. We'll take up Sanskrit or vegetarianism and find a new life."

"Thank you. Except you're making stupid jokes again and I don't want a new life. I want the life I always wanted."

"Which one?"

"What?"

"Which one? How far back does the dream extend? It wouldn't give us any more comfort, unless we got back to when he was a little baby. Before he knew he was leaving us. I mean, he tried to run away from home when he was *five*. Remember?"

"He asked me to give him money for the bus."

"So what I'm getting—don't do that."

"I'm not."

"Not what?"

"Not crying."

"Oh."

"Mac, will you shut up? Keep talking, what were you saying?"

"I don't know. I don't remember."

"Mac."

"Yes. Right. Yes. Ah—even if he came home it would be this. Something like this. He really wouldn't be there. He can't be. From now on, he visits. If we're lucky. He'll need help, he'll need money, he'll even want comfort, maybe even advice some time. But he'll say 'your apartment' instead of 'our apartment' and he'll come to see us as a guest. I'm sorry. I really think so, though."

"So we should make ourselves change now?"

"We don't have to make ourselves change. Except it'll happen because we'll die if we don't. But everything else is changed now. That's what I'm saying."

"Well I think we'll die if we do. Even if we do change. Because what you're talking about is that suddenly we're a hundred years old."

"No. I'm saying we have to figure out some ways to be a little less vulnerable."

"And what *I'm* saying is that you're lying to me, Mac. You're the biggest crybaby in the world where he's involved. I used to watch you check to see if he was breathing while he slept. Every night."

"Twenty years ago."

"Now. Now."

"No."

"No?"

"There's still you and me. That's what I mean. We're still—my life still depends on you. Period. We're still you and me."

She drank the whole glass of ale as if she chugged it at a fraternity party, and she wiped her lip with the edge of her hand. She burped and smiled and covered her mouth. She said "You're still jealous."

"Of my own son? What about that checking every night on the breathing business? Come on, Claire. You've watched me chew myself so long over him—"

"I'm not saying you don't love him. Didn't love him all the time. I never knew a man who was a more worried father than you. No, you loved him—"

"Love him."

"Yes. Yes. But you're frightened now. Like me. And you want me to say what the hell, he's left us, he's gone, and now we'll take care of each other and make everything all right."

"I want that?"

"Yes."

"Yes."

And in their room above the juke box that hammered the walls, while Claire took her yellow linen dress off, he took off his gray-and-white striped seersucker coat, transferred his money from his billfold to his trousers pocket, left the billfold on the bureau, tucked in his blue long-sleeved shirt, went to the door, said "I love you and I love Harry. I think I can find where he lives. I love you. Go to sleep" and went out while she was still trapped in her clothes. He ran along the carpeted hall to the stairs and ran down. In the lobby he walked, panting, then stopped in front of a girl with an American Indian's beakish face who, dressed in a man's undershirt and white-painter's coveralls, no shoes, slumped in a green vinyl easy chair and stared at him as he sucked for his breath.

He said "Can you tell me how to get to the part of the lake where a lot of the students live? I used to know how, but I forget."

She pointed to her right and said "As far as you can go."

So he was driving down a long residential block off the main street which had the town's stop light, and he had the radio loud, he was banging on the gear lever and nodding his head as he drove with the windows open slowly past Greek revival houses and stone colonial box shapes, the music scraping over house after house like a long shadow. When he reached a small iron bridge and the golf course began, he turned the radio off and slowed and stopped. Wind through the pines near a wide trap hushed, the green on either side of the road went on, and there were only trees and grass and the wind's sound. Behind him, in the light of the street lamps, most of the houses were dark on the first floor, lit on the second by the yellow lighting of a single room or the blue of television. Out of sight beyond a curve, but reflected on the clapboard siding of a Cape Cod house, the stop light pumped up red. He left the radio off, put the car in gear, rolled on about an eighth of a mile, then stopped again where the road bent to the left, around the golf course, but also, straight in front, turned into one dirt lane. "As far as you can go" he said. He turned left and went faster.

There were fewer old houses, more new ones, long quarter miles of forest where no one seemed to live, and then, around an S-curve that almost tipped him off the road where a small sign said RESERVOIR, he came to the lake, which was invisible. The air was colder, the mist lay out on the lawns which went back to little houses and cabins and dark shapes of bathing suits that shook like skin on the clotheslines. The lake was hidden behind the buildings he passed, his headlights reflecting the mist before him and not lighting up the road, but he continued to roll around the changing shape of the lake, looking to his left, where the cabins were, and not to his right, where the forest came down to the road and formed a wall.

The music drew him onto gravel and stone chunks, then mud. The little house was a hundred shapes, it had been built and rebuilt, added onto, renovated, patched, damaged, almost fixed; its shadow changed as the car pulled in, as it stopped, as the headlamps bounced, as the lights went off, as he opened the car door and closed it gently, as he moved. The music was constant, Ornette Coleman's frantic impacting. The lake behind the house made the sounds of sea on a stone shore. The mist lay as high as his ankles.

And in the dark house—there was a light inside and the mist outside responded to it, but the light itself was unseen— a girl said "Because it isn't just a question of responsibility. I mean, nothing could be as simple as that. Otherwise everyone would be *wrong* all the time."

A male voice said "But everyone is, Stace. That's the trick of it."

And the girl's voice said "What an incredible drunk you are."

Harry's voice said "A jughead."

Somebody else said "Jughead" and everyone laughed in the music.

Mac walked along the side of the house, touching its peeling clapboard that smelled of garbage and gas, the lake water, trees. He rubbed at his chest. And then the house stopped and a short rocky lawn went down to the lake, which was covered with mist that glowed beneath the low glowing clouds.

Stacy's voice said "You can all hear about my paper now."

Harry said "You'll do it, you'll do it. Leave it alone for a while, relax."

Stacy said "If I don't do it by graduation time, I get an Incomplete. Which I won't do, which I will get, because tomorrow's graduation. Which means I have a month to get the paper in and get a grade and pass or else I win the big F. Which means I'm going to have to talk all summer about failure with my father."

Harry said "Your father's a nice guy."

She said "I'm a nice guy too. Everybody's a nice guy. But he'll talk all summer about it, I mean it. And what am I going to say all summer about failing?"

Harry said "How about *whoops?*" Everyone laughed and Harry said "Hey, how about instead, how about a little hit of Mister Dan's magical pass, huh? Small hit? Could I interest anybody here in a hit of reefer-pass?"

Jughead said, in a deeper bass, "Reefer past, reefer future: I am the Ghost of Reefer Past."

The girl said "I can't *say* anything, I never can. I sit down and I start to write about the French Revolution and I—get scared, I get stopped, I can't get anything committed onto the page. I can't get it *out* of me. I might even know something to say, but I can't say it."

Harry said "So this summer we'll write it."

There was silence under the music and then the music stopped; there were shufflings and clicks, the same music began again, "Lonely Woman."

Harry's voice said "What?"

Jughead said "If you guys need a discussion here, I think I'm gonna do a little wake-up action on Roberta and cut out if you think maybe we—"

"No!" Harry said. "No, no problem, stick around. I just never heard that from Stacy before is all. No. Hey, Stace?"

She said "All I said was I think I'm living at home this summer. My father's doing some string-pulling on the social work people in Toms River and I think I'm trying to tutor a little kid or something. Something like that. Because if I can get a little kid to learn, maybe I can get *me* to learn. I don't know. I've been in college three years and I can't *do* anything. I can't write a ten-page paper. You can't be not able to do anything all the time, Harry, right?"

"Wow. Wow, we had these other plans, I thought."

"Well."

"About living here and I was painting houses and you were working at the Inn and we had this—"

"I can't. I don't think I can."

Jughead said "Me 'n Roberta are going. Catch you tomorrow, folks. Call us up if you want us, all right? Gotta go, gotta go, gotta go—hey, broad! Hey. Wake up, broad!"

Harry said "Everyone's going away, what *is* this?"

Jughead said "Say bye-bye, broad." There were mumblings and then Jughead's voice—"Hey: be good to you guys, you guys"—and then the music going on.

Then Stacy said "Is it so bad?"

"Well you were scared to tell me."

"That's what I'm saying—I'm scared to tell anything. That's why I have to do something this summer. Soon. I'm supposed to be a grownup soon, Harry. I'm getting scared."

"Well, listen, Stace, maybe I should work in the city or something and take the train down on week ends."

"Maybe."

"Oh."

"Well *I* don't know, Harry. I don't know."

"Yeah. Maybe I won't do that then."

"Well."

"Yeah."

Then Jughead's voice said "Stop the action, folks, there's a big black car outside. Anybody here call the *Polizei?*"

Stacy said "A raid?"

"Flush" Harry said. "Run and jump and flush. I'll do this piece and there's some in the cushion, over there. *There.*"

Roberta said "You really know how to have a party, Harry—breakups and break-ins and everything."

Mac went through the doorway to the back porch, past a

55

wet life preserver and towels, to a screen door, through it, through a small dark room that smelled of shampoo, into the room where Harry—lighted by candles that sizzled and threw darkness onto the barely lighted walls—was standing with a short bearded boy in glasses before a sofa that was propped by books. He said "Harry? It's me—I'm sorry. It's me."

A tall girl in a dark tee shirt and dungarees and high-heeled shoes, her body big and sexy, her pale face smooth and small as a young child's, pointed to him from a far doorway. A short and very skinny girl with long black hair said "Where's your warrant? Where's your warrant? Toby, doesn't he have to have a warrant?"

The short boy with Jughead's voice said "Right, sir. Would you happen to have a search warrant with you?"

Mac said "I don't have to have one. If I were a policeman, I mean, I wouldn't have to have one. Not for drugs, I don't think. I'm not a policeman, it's worse than that, I'm afraid."

Harry said "It's my father."

Mac said "I'm sorry. Did I hear someone talk about wine?"

Harry said "Why didn't you knock on the door, Dad?"

"Could I have a drink of that wine with you?"

"I mean, why'd you sneak around?"

Toby said "Like before, friends, we're gone. Sir? Nice to meet you, glad you're not a cop, and have a good evening, good night. Good night."

Stacy, across the room, said "Harry, I'll call you."

And Harry, bare-chested in blue jeans, his chest and stomach flat but flabby, said "You won't leave tomorrow?"

She said "I won't leave tomorrow."

"But you're taking off?"

"I'll call." She smiled toward Mac, but the smile didn't reach across the room, and then she followed Toby and Roberta. Harry pushed a lever on the record player next to a wicker chair and the room became silent. He said "She'll call. Like she'll do her term paper on the French fucking Revolution she'll call. Thanks, Dad."

"Harry—"

"Thanks so much for coming out." Harry kicked at a small pipe on the dark rug that merged with the scuffed unpainted floorboards and it rattled in a wobbly circle where it had been. He said "No. No. I don't mean you couldn't come out."

"I didn't mean to ruin anything. I don't want to ruin anything. She's a beautiful girl."

"She's very immature" Harry said.

Their shadows on the shadowed walls. Their perspiration, in bubbles, greasy, on their foreheads; in long streaks on Harry's chest and sides; in dark wide smears on Mac's blue shirt. Their breathing: shallow; caught, released, as if the act were up to them. Their motion: careful, clumsy, slow. The sounds of water on the stones outside. The cabin's own noises of wood contracting, of branches on the clapboard, of breeze against a window screen, of gas leaking slowly away. The darkness that they floated in. The pulse in Harry's throat. The trapped jumping eye in Mac's head.

Ornette Coleman, then, concussing out of the record player as Harry pushed the play lever. Mac stepping back as if the music burned his skin. Harry slowly pushing on the lever again to turn the music off, saying "I'm not at war with you."

Mac said "I'm not at war with you either."

"But don't you hate this?"

"Harry, we have to talk tonight."

Harry said "I wouldn't know how."

"Then *I'll* talk."

Harry sat down on the floor and crossed his legs in front of him. He put his hands on his knees and waited. He said "Okay. That's good. Okay."

And Mac said "I don't know what to say."

A telephone ringing against a wall. Another room in the house ringing into where they were. It burred and burred on wood or plaster, and Harry and Mac sat still. Then Mac drank more wine from the half-gallon bottle of mountain red—hoisting the jug and drinking, spilling the cheap wine onto his throat and shirt, not wiping his mouth—and Harry, watching, sipped wine from his white Woolworth's coffee mug and put it down when Mac replaced the wine jug on the floor. The telephone rang, and Mac said "That isn't Stacy, it's Mother. She must be worried. She must be completely terrified."

"We'll let it ring?"

"What would we tell her if we answered it?"

Harry said "We could tell her we're talking."

"We haven't said anything yet."

"Well that's how we're talking right now."

Mac put the wine jug down and said "Do you love Stacy? That's dumb. You love her, correct?"

Harry sipped and said "I love everyone, Dad."

"Do you love me?"

"Don't ask that kind of question."

"Do you love Mother?"

"Please?"

Mac said "I guess it's pretty clear I don't know what I'm doing here. You know, I once lived out at this lake. After I graduated. I taught here once, for a few months, and I lived out here. Maybe where you live, I couldn't tell now. And now I don't know what I'm doing here."

"Well, you came."

"You wanted me to?"

"Dad, there's nothing that simple—nothing I know is as simple as that, yes or no. It just feels like I'm glad you're sitting here getting lushed."

"You think I'm getting drunk?"

"I wish you would."

"Why?"

"So we could both be out of control."

"Oh, we are."

Mac stood in the doorway between the living room with its candles and the kitchen from which the garbage smell came, the jug held in the crook of his finger by its glass loop. Harry lay on the floor with his head back on the cushion of the low wicker chair. He looked at the ceiling, then at Mac, and Mac kept looking at him. He shook his head and said "I'm not feeling this wine."

Harry said "I wish I had something stronger for you."

"I don't think I'd feel anything stronger."

"You're too upset?"

"I didn't come out here to put my feelings on you."

"Why'd you come, then?"

"I got in the car and came. I don't know."

"I like that."

"Impulse?"

"Things you can't explain."

"Why?"

"I can't *explain,* Dad." Mac giggled and stopped, kept grinning, laughed out hard—a cough from the back of his throat—and shook his head, coughed smaller, stopped. Harry said "My ass, you don't feel it."

"No, no, it isn't the wine. I don't know what it is. But I know what you mean. Mother and I walked around all night tonight—"

"I'm really sorry that happened. See, I had this feeling about Stacy."

"Please. Please, that was fine. We walked all over the campus—it's a beautiful campus, I'm glad you went here."

"Yes."

"I hope you're not *sorry* you went here."

"I'm very grateful to you and Mom."

"Dammit, Harry! That's not what I'm saying, what I want. I'm saying, all I'm saying is we walked, all right? We walked and we talked to each other a good deal and we—well, you know we're concerned for how you feel. We were talking about you. And we kept on saying things *we* didn't understand either. Things we couldn't explain. We kept trying to explain them anyway. And we couldn't. Which made us more upset. And you wiped away the whole problem with what you said. I forget. Not the whole problem—dammit, I don't even know what the problem *is*. I don't know: I liked what you said." He drank more wine and wiped his mouth, let the jug dangle and said "What were we talking about, at the start of all that?"

Harry, looking at a candle in a saucer near the door by which Mac had come in, said "You're a nice man, Dad."

"Oh."

"What?"

"Thank you. You're a nice man too."

Harry looked at the candle and clasped his hands across his stomach. "So what should we do, Dad?"

Mac's face opened and he opened his mouth, shook his head. He waited, then said "What would you like to do?"

Harry lifted his shoulders. He said "Why don't we burst into song?"

With the moonlight less clear than before, the luminous clouds a distant silver, Harry, naked—tall and recently thin, built like a stretched-out boy except for the thick wobbling penis—walking down the scrubby grass to the lake. Mac walked behind him, tentative and on his toes, naked too—fatter, less comfortable with his flesh—and he followed Harry into the water. Harry swam out and so did Mac, Harry with long wild strokes, Mac with a smoother crawl. Then Mac lengthened the stroke and kicked out and the black water foamed behind him as he reached his son, whose flesh was phosphorescent in its motion. Mac went on ahead, then stopped to tread. He panted and coughed. Harry stopped too and they paddled in place, gasping, wiping their noses and eyes. Harry started a stroke and Mac jumped into his and went ahead. Harry stopped and then Mac did. Then Harry stroked and Mac leaped into his crawl and went ahead to stop and wait. They treaded, wiping their faces, hissing in and out.

Mac said "I once. Almost Drowned. A long time ago."

"What'd you think about?"

"Sex. Completely pornographic. I thought. About all the girls I ever knew. Then I thought. About the one. Who liked me the most."

"And then?"

"I got saved. I was sorry. To be saved. They pulled me out. With a giant erection. Everybody laughed. While they made me. Breathe."

"You telling me something?"

"No. Swimming back." He leaned and kicked and swam slowly back, then turned onto his side and stroked gently, not making much progress but going toward the shore, and Harry lunged into his grabbing crawl and followed.

At the harsh edge they sat on stones and panted. Mac rubbing at his chest insistently, Harry in his Indian squat. After a while Harry said "What should we do, Dad?"

"That drowning business—I wasn't trying to tell you anything. I'm not that unsubtle. Well I'm not that subtle either. But I'm not trying to get anything across. Because I don't know what I want you to know. I was remembering. But it was a good thing to remember."

"Yeah, but what should we do?"

"You think we have to do something?"

"It feels like it."

"Do you have any ideas?"

"I want to stay here this summer."

"Right."

"I don't want you and Mom to come up. Okay?"

"Yes."

"I'll send you a post card or something."

"Right."

"And we'll see."

"See what?"

"I don't know."

"That girlfriend of mine, Harry. I just wanted you to know: I really wanted to marry her, or stay with her—something permanent. I wanted to keep her."

"Keeping is big medicine, isn't it?"

"And she wouldn't let me. It was bad all the time with her and it was worse after that. When she decided we ought to forget it."

"Why did you want me to know that? Because of Stacy?"

"No, I'm sorry—I forgot about her for a minute. No. I wasn't thinking of girls at all, I guess. I just wanted you to know it hurt while I had it. While we were together. When I thought we would *stay* together."

"I don't know what you're saying, Dad."

Mac stood and went farther up on the grass, wiping silt and sand from his feet, smearing his feet as if he were walking, but staying where he was. The fat on his body shook and he turned partly away from Harry as if he were ashamed. He said "That even if I don't know what this means"—he waved his hand at the darkness behind them, the darkness in front. "What you're really doing. I know it could be plenty damned easier. And I don't hate what you're doing. I don't hate you."

Harry sat at the edge of the unbounded lake and didn't turn around. He said "We'll see."

In the silent Inn, in their black room, sitting on the bed and waiting for Claire to waken, he rubbed at his chest. She coughed and sat straight, a child he'd wakened from dreaming. She said "I wasn't asleep. I was waiting."

"It's all right" he said.

"I wasn't asleep."

"Harry's back at his place, he's asleep. Everything's fine."

"Why did you run off like that? Without considering me? Do you think you're a hero? Is that it? Were you trying to earn *medals,* Mac? I was *frantic.* I called everyone I could think of and I would have called the police—except for what Harry would have said. I was terrified! That kind of thing is so incon*sid*erate!" She turned and lay on her side, looking away. "You know what I think of that kind of behavior."

"Yes, I do."

"So why did you do it?"
"I was getting rid of a rival."
"What?"
"Just a joke. Another joke."

The telephone rang and she cleared her throat. Mac went in the darkness to lift the receiver, listen, nod, still rubbing at his chest. He nodded again as he held the receiver away from his ear. It ticked once, and he hung it up.

Claire said "That was Harry."
"He said to have a safe trip home."

LET
SLIP
THE DOGS
OF PEACE
1961

I think again tonight of Rogovin—I say "So long." I say "Good-by, go win the war"—and, springing alone on my bed which goes *ratchet, ratchet,* I hear the same steel music that the elevator played in U.S.A.F.E.E.S., New York, the khaki corridors and slamming doors, when Rogovin and all of us were pale and much too loud or soft, too young, waiting for the Army doctors on the long floor. The *ratchet* of the hung cage descending in the well of those national stairs as morning creaked in.

There was a smell of last night's floorwash of ammonia and a yellow on the ceiling from the underpowered bulbs. A small man in military blue brought coffee in a paper cup and watched us watch him drink it. We sat on wooden benches, staring as the coffee disappeared, we waited to begin. He rubbed at his freckled face and asked a hundred of us "How early can you get, huh?" and listened to the answer coming in: shoes on the vinyl floor, a scrape of elbows over wood, low hum. He smiled and lit a cigarette, sat on a table and swung his legs; he spread his short arms out and told us "Men—"

A voice from a door to his left said "No smoking."

"That," said the one with the cigarette, "is how the Army works." He pointed to the side door and a soldier, not much more than five feet of vertical creases, green, and said "They treat you that way in the Army. You should join the Air Force."

The one at the door said "And drop things on civilians?" He put one finger into the top of his crewcut and told us "Smoke your sleepy civilian heads right off, men, puff it up, everybody light *up:* work that cancer up big, blow it up big now, you can stay home a while, get the pick of the women. *Light* up."

"Men," said the one in blue, "there will be no smoking on the examination floor due to military requirements and not to mention fire laws." He looked straight into us and said "I would advise you now to grab a smoke while there is still time." He pointed again to the side door, now empty, waved his legs back and forth and smiled and told us "I'll take care of them."

Smoke went up like flies in a field, and boys in shiny pants uninjuredly limped, waved their heads, shouted over the benches and the smoke: "Hey Petronis. Petronis! Hey: Petronis!"

"Ho Duke."

"Petronis, you joined up?"

"Hey Duke!"

"You joining up?"

"Scratch my titty, will ya?"

"You joining up?"

"Where's the kid?"

"They called you down, huh?"

"The kid get a letter too?"

"Who?"

"Hey, I'll see ya later, right? Duke? Catch me after, right?"

"Yo."

There were grins and the sounds of furniture moving, the

ratchet of the elevator outside in the center of the double stairs. The fat boy next to me said "Here goes matches."

"Excuse me?"

"Matches. You looked like you needed some matches. You want some?"

"Oh! Yeah, yes, thank you. Thanks a lot."

"Watch it" he said. "Forget it, here we go, kill the smoke."

Another short one in green stood at the front of the room and took a stack of brown envelopes from the one in blue and then called out "No smoking. Butts out. Let's go." I stood up and he said "Take your seat. Resume seating, there. Move only upon my clearly feasible command."

The fat boy next to me said "Duffy of San Quentin, that's old Duffy, I had him here the last time."

I put my cigarette gently on the floor and stepped on it while the fat boy said "This your first time?" I nodded. "Start working up a piss" he said. "And don't talk too much, they hate it when you talk. But scared? Look scared. Everybody's your mother right away, just look scared and try not talking. Okay? I know the bit, the whole thing, watch the way I do it."

"I don't want to go to Berlin" I said.

He said "Are you four-eff or something?"

"I don't know. Maybe on account of my eyes?"

"No, I mean the class, you know. What did they classify you?"

"Two-ess."

"You go to school?"

"No. But I used to."

"Maybe you're still okay. Can you lie? Tell 'em you still go?"

"I don't know. I don't know. They told me to come, they must want me."

"Listen, they want everybody." His fat fingers, knuckles black, ran figures on the air. "You have to lie" he said. "It's okay, you don't have to sweat a thing, I'll tell you who to talk to at the end, but do the bit, they don't listen till you do the bit.

You just hang in and work the piss up, run through the zoo parade, I'll tell you who to talk to later."

"Hey-oh!" Duffy told us what it meant. "Re-sume silence, men, we are about to commence."

"I had him the two last times" the fat boy whispered. "Every time they marked me down fat, he personally right away got pissed off. He's very serious about being fat." He pushed the sleeve of my loden coat and I turned around. "You're not too fat, you know? You maybe can't make it that way, but it's okay. Don't talk, look scared: I'll get you through the bit."

I looked toward Duffy and the fat boy pushed my sleeve and said "My name is Rogovin" and pointed, hand held low, to the front of the room. "Catch the Beast of Budapest bit."

"And also to help you" Duffy was telling us. "Far from filling a local quota, and far from making sure nobody succeeds in the evadance of his righteous civil duties, and not only to carry on the law—that's, ah, not the only *thing*." He looked me in the eye. I closed my eyes. "Do you see what I'm saying, men? You have to get this now: far from, ah. That, *all* of that now. Some of you are *sick*.

"Now. Look around you." The benches rocked, squeaked, as the boys looked around, as the elevator ran in place, as the cold floor trembled to the subway underneath. "Some of you are *sick* and you don't know that. But I'm telling you the justifiable truth. We know, we know how to find out and *cure* you so that you *can* take all of your places, every last man in this room. In the service of your country. That's right. If you have cancer and don't know it, we have ways of finding out, and so forth, heart disease, cataracts, obesity, the whole gambit of physical unrest.

"So. Why are you here so early on a Monday morning?" He held his hand spread wide for us to see. *"First,* duty to your country." The thumb curled away. "Following of the laws, ah, you must follow." His index finger knuckled in behind the thumb. "Self-help if you are ignorant of one or more diseases you may or may not have." His middle finger folded on the thumb and the remaining two tilted forward, wavered, dropped then into a hanging fist we watched.

"*There*fore, you will come to this desk upon receiving your name to pick up *one* envelope of that same name and *two* New York City subway tokens constituting carfare reimbursement traveling to and from this area. *Pro*ceed."

And then the two of them, Duffy in green, the other one in blue, called off a hundred names and got a hundred answers, and one hundred envelopes, physical questionnaires clipped outside, were carried on the vinyl floors through green hallways to the double stairs that climbed around the elevator shaft like wide dark marble vines. I patted for my wallet, then my keys, my loose change. I pulled at the hem of my coat, pushed the clip on the envelope, sniffed at the cold recollection of dust that the lights gave.

Rogovin said "Don't *sweat* it, I'm telling you this is no sweat. I had this bit four times, no pain."

"Just work the piss up" I said.

"Listen, everybody goes through here, they don't care about you, they just want to get your ass out. They're in a hurry. No sweat. Look: you're a doctor. Right? You come here and you stand up there all day and look for piles. Right? How particular can you get? Right? And you don't even have to *be* here if you lie right. Jesus, don't worry. Look: you want some help, I'll get you through, what the hell. Okay?"

"Is the whole thing naked?"

"Huh?"

"When do we strip?"

"Oh, they get you down pretty fast. Why? You like that stuff?"

"I mean, do they let you keep anything on?"

"What, you don't like it? Or you do?"

"I don't care."

"No, I mean it."

"I don't care."

"You ever read a book called *Blonde Girl's Buff?* Detective book? They put it out in paperback."

"No, I—"

"I read a lot. I don't have too much time, you know, so I can't see dying in front of the TV, I read. Not like you, but I like it."

"No, no, I read detective books, I like them a lot."

"Ever read *Blonde Girl's Buff?*"

"No, I think I missed it. I'll check in the library, though, you really recommend it, huh?"

"I don't remember who wrote the thing but they got a strip scene in there? Jesus."

On the third floor we walked along a balcony that surrounded the stairs that surrounded the elevator. No one gave directions in that unlighted landing, but Rogovin chanted off the signs on doors we passed, calling "Visual Examination," "Feet," "Audial Perception," until I breathed to the rhythm of his words. Whoever led us walked into a small room filled with chairs. A short black man in green watched us through black-rimmed glasses and, when everyone had sat, he snapped his fingers.

"That," he called in a high voice, "is the sound you will respond to henceforth. Which means from now on in case you are not an educated individual." He snapped again. "That means look at me and shape up and do as I tell you." He snapped again. "Do not fail to hang onto every syllable I form."

He snapped. "I am coming among you to leave off Army-issue pencils and I will not release this muster until each pencil is accounted for." He came among us and left us pencils and a smell of coffee that made my mouth dry, the sides of my head feel thin. Then he snapped his fingers and we focused on his mouth, the pale lips cutting with precision in the small cold room of chairs. "You are now to fill in the questionnaire printed in brown ink. Noting the large chart on the blackboard, using it as an example, you should have no difficulty, but I will help you free civilians through the first few questions starting *now.*"

He snapped. "Name first." He folded his hands in front of his crotch and sighed. "Who does not know his name?" He looked at the back of the room and nodded his head. "Your name is *sheeit*. Is that clear? Good. Who else?"

He snapped. "Number two calls for race. The Army does not care for the fine distinctions and you will therefore enter Caucasian if you are white, Negro if you are black, Mongolian if you are tan, and Toilethead if you are either pinto or palomino."

Snap: "Respiratory defects will be answered as follows." Snap: "T.B. will be treated in the following way." Snap: "On the bed-wetting question. I am certain some of you will have to answer yes." Snap: "Spitting blood." Snap: "Motion sickness." Snap: "Night sweats." Snap: "Upset stomach." Snap: "Urinary pain." Snap: "Any kind of ulcer." Snap: "Recent loss of weight." Snap: "Dizziness at heights." Snap: "Claustrophobic symptoms." Snap: "Frequency of headaches." Snap: "Numbness of the limbs." Snap: "Scarlet fever." Snap: "Jaundice." Snap: "Smallpox." Snap: "Measles." Snap: "Mumps." Snap: "Fits." Snap: "Death."

Snap: "I said who here has had a case of *death?*" The glasses looked us over. "You ain't dead, civilians. You are just lying down on this very easy job and *playing* dead. And"—snap—"you will therefore sign your names at the bottom of the sheet and give me those Army-issue pencils back and leave me be." He snapped his fingers and came among us and we gave him the pencils to count and he waved us away.

Rogovin walked up beside me as the group—was someone leading us, did someone know where all of us were meant to be each time we moved?—was eaten by the darkness at the doorway of the small room with chairs. "I tell you?" he said. "Like a machine, they never know you're there. I'll get you through."

"Would you carry aspirin around with you? Would you have some aspirin I could borrow?"

"You don't have a headache" he said.

"Now, that's a relief, Rogovin. You know, I was walking

up these stairs and thinking I had this ferocious headache until you told me how silly I was to believe it. That's my trouble: every time my head hurts I think I have a headache. I know it's silly. Thank you for your help."

"You're only scared."

"What?"

"You're scared. It's all it is, I had it the first time through. You know, you keep feeling like all of a sudden they grab you and put you in a truck and put you in a plane and you're in a land war in Europe and your family's home crying, it's all over, you're finished. Right? Okay. Except there's enough guys to get without you and me and they don't *want* you. Right? So make believe you're not here."

The elevator, clanging in the shaft, going *ratchet* through the stairs, was on the fifth floor waiting for us, and there was Duffy too, in green. "Form up, men. Two lines, coats and ties off, right sleeves rolled high, coats and envelopes carried in the left hand, *pro*ceed."

We walked around a large room filled with chairs and cots, the room so brightly lighted that I squinted, squeezed my temples, looked from the light wooden floor and folding furniture to the wall, where a black sign said in white letters THE MEN ARE IN THE ARMY NOW/THEY'RE BUSY/DO YOU WANT TO WAIT 'TIL THEY COME BACK?/OR WOULD YOU LIKE TO JOIN THEM?/. . . IF YOU'RE GOOD ENOUGH. I looked back to the room, at the cloth partition against the rear wall. A voice behind the cloth called "Okay, Ralph" and Duffy waved the boys at the head of the line inside.

Someone behind me whispered "Confession" but no one laughed, no one spoke—except Rogovin, who told us "Little blood sample, no sweat, you don't even feel it."

A tall blond boy in front of us, white shirt transparent with sweat, slapped at a chair and it collapsed against the pale glossy floor like a stiff man dying.

Duffy called out "You like to break things up? We can use you in a country I heard about. You pick that blackass chair up and stand it upright parallel to the ground and *re*sist from fooling around this-*forth*. Is that clear?"

The blond boy rubbed his hands on his thighs and said "I wasn't tooling around."

The line moved up toward the partition. Rogovin tripped and fell like a filled-in gong, saying, as he swayed on his knees like something rung, "He's humping the whole thing up, he's humping it up."

Duffy was there, telling the blond boy "I said *fooling,* sonny, I didn't say a word about *tooling,* and you will *not* instigate words in my mouth. Co-*rect?*"

The line moved up. The boy said "No sir."

"Because that places you in dire, ah—doesn't it?"

"Yes, sir."

"Okay. Now. What's your complaint?"

"Sir."

"The problem. We're running out of *time*. What's the problem with you?"

A boy in loose brown pants walked from the other side of the curtain, his forearm at a right angle to his biceps, a vial of dark blood in his high hand. He rolled his eyes back and smiled at the line. The line moved up. The blond boy said "Can I get out of giving blood, sir? I can't do it, I get sick, I faint."

"You a hermaphrodite? You bleed too much? You carry a card?"

"No, sir. I get sick."

"From *what?*"

"From giving blood." The line moved up. "I get sick when I do it, I can't do it."

"Learn it, it's a thing you have to know how to do."

"I'm not gonna do it, sir." The blond boy spread his feet and Duffy tensed on his toes, crouched. "I'm not. I'll get a letter from a doctor, I'll come back this afternoon but I'm not gonna do it. Please."

"Hump the whole thing" Rogovin said. "Screw us all. You see?" he said to me. *"That's* what gets you screwed, when you bunch the line up like that, they hate it. That's when the shit flies, you watch."

Duffy, still on his toes, moved in. Someone behind the partition shouted "Ralph" and someone at the end of the line sneezed.

Duffy called "I got him, it's okay, I got him" and he moved in low, chopped his foot down with a long stride and screaming "Ki-*ah!*" reached for the blond boy's arm, pivoted, fell with the arm on his chest, the boy on top of the arm, and was pinned.

"We're finished" Rogovin said. "This is it, we'll stay here all the humping *night*."

The boy stood up and said to Duffy, on the floor, "You did that, I didn't touch you. Everybody saw that. You tried to throw me, I didn't touch you with my hand, you're not allowed to kick us around."

Duffy, on his back, looked at the boy, moved one leg slowly, then planted it in the blond boy's crotch. The boy went down like a folding chair and Duffy pulled him up by the hair, held him until his eyes opened, then pushed him to his knees, held the boy's head down. "Move the line" he said, panting with clear regularity, "continue this operation rapidly."

"That's it" Rogovin said. "We're screwed."

I walked behind the partition, sat on a stool and closed my eyes. "You can open them" someone said, but I kept them shut and, by the time I had whinnied in pain, I was walking to a chair, holding my blood aloft, watching the line move up and the boys come out with vials and others, rested, handing in their blood to a white-coated, hairless man near the door of the bright room that smelled somehow of vinegar, then leaving with the fist clenched at right angle to the biceps, Duffy and the soundless gagging boy behind them on their knees. I walked slowly to turn in my blood, left slowly for the dark and chilly hall around the elevator shaft, moved slowly with the others so that Rogovin, his shallow breath, swift phrases, could catch up and caution me that we were not screwed.

He gasped behind me as we all moved down the stairs, he said "We're screwed."

Someone in the darkness chattered "Hey, the big guy says we go get screwed now." Someone else laughed. No one answered anybody, then, and the elevator winched up slowly from below us.

"Rogovin," I whispered, "how am I supposed to stay calm if you go pessimistic on me?"

"You can talk out loud" he said. "It's only when the rough guys are here, you know, drop it down a little so it's like you're not around."

"Rogovin, you sounded panicky before, twenty seconds ago, up on the top of the stairs. What's the story? Do we cool this thing or not? When should I start to worry?"

"Listen, look: don't sweat it. Okay? I know the bit, in-out, they give us lunch, we go home. Don't sweat it, I'll show you what to do."

"Don't sweat it."

"You got it, that's the way to do it. Now you're okay."

"Where do we have to go now?"

"Third floor, physical exam, men's underpants, lady's bloomers, underarm deodorants, and fine furniture, step down, keep your—"

"This is where we strip?"

"Well they got to *look* at you, right? How are they supposed to check us out? Right? Yeah, we take it off. They got the coldest damn floors."

"Great, I love this, I really love this because I like to suffer so much. Rogovin? Rogovin? What'd you mean, before, upstairs? When you said we were screwed? Did you mean that?"

"I can't do a goddam thing if you panic, you know? Panic, and I can't help. You never heard a figure of speech before?"

"All the time."

"Okay: so?"

"So I don't sweat it."

"Rah-*toe*."

Someone turned the lights on, then, and the little room, its

74

many doors, barred window, wall of hooks, turned on as if a bulb beneath each board were hot and friendly, and as if the day were given, now, a second chance. Of course Duffy was there, smoking, and so half of us dived for cigarettes while he watched us.

"There is no smoking on the examination floor."

"This is a very unconventional day" Rogovin said. "I never saw him here."

The big black man in a white jacket rubbed his chin and looked at Duffy and, dropping his voice again, said—his chest said—"We do not smoke on the test floor."

"I would love to stand up, then, and finish my cigarette" came from high-heeled boots and light blue corduroys and a thin wrist wrapped in a silver chain. "May I do that?"

The big man looked at Duffy and Duffy looked at the barred window. The black man looked at the floor and said "You extinguish the *fag* right now or I find a fire extinguisher and freeze your lips with it to the back of your highly swish mouth in about two seconds, son. Do you see what I mean?"

He backed into a door marked MEN and, when it was half open, faced into it and went inside. The wrist in silver chain called out "I hope you drown, you big black bastard. I'm a civ*il*ian."

"Death" said Rogovin. "The queers are screwing us all."

"Men" said Duffy. "This has not started out well and I want to give you some advice: shape up. That's all. Get squared away and shape up. Is that *clear?*

"Now. Upon my clearly feasible command you will derobe to one sock and undershorts, leaving all other garments at the window to your direct right, carrying all jewelry and valuables in one sock gripped in the right hand and brown envelope in the left. Upon turning in suitable garments you will proceed. Through *that* door to get weighed and there will be no loud jokes about getting weighed. Good.

"*Is* that clear? As soon as the gentleman at the scales informs you to proceed, you will do so, refraining from all conversation and jesting in order to render yourselves open to sudden commands or pertaining questions. Okay. Good. *Pro*ceed derobing."

"Rogovin" I said, but he was gone, prying shoes off with his toes, simultaneously unbuttoning his shirt and yanking at his belt: a thick-fleshed flower throbbing pale from its woolen bud. By the time I had my trousers on a hanger taken from the wall, Rogovin was female breasts unhaired and yellow rolls of goose bump hung on the whiteness of his brief shorts. A smell of skin went up like smoke and nearly naked Rogovin, racing past the fag in shorts of blue, was diving, almost, through the fog to turn his clothing in and be discovered as obese.

I walked slowly to the clothing window, flexing my feet carefully, trying not to shake or roll, then went into the weighing room. Rogovin was on the giant scales when I got there, and I hoisted my sock to say hello, but he was looking at the dial and saying "That thing's off, sir. Could that be possible? That it goes under? You know, shows too little, something like that? I don't care—you know, but I got obesed last time and I didn't cut down on a thing. You know? I *know* that."

The Puerto Rican in a white jacket wrote on Rogovin's form, gave it back to him, said "You too fat, but you not too too fat. We cahn use you foh a spay tahir if a truck break down. Nayxt. Good-by, too-fat, hey I say *nayxt*."

"No," Rogovin said, "you can't just do it that way." He climbed from the scales, reading his form, shaking his head. "No" he said. "Look, who's your boss? Can I talk to the C.O.? Somebody like that? You can't just do that."

The Puerto Rican waved his hand and Rogovin stood, legs apart, sock in one hand, the paper, with the weight of his days, in the other, moving hand. Then Rogovin went into a narrow green corridor, and the line moved, and I moved up to the scales, and onto them, and off, and into the green-walled hall where a line moved slowly toward a room that generated clicks and echoes of steel, the bellow of an outraged mouth.

The big black man stood next to a tank that had glass doors on it, and knobs, levers that looked cocked. "Card to me, chin on the face-rest, chest against the lens."

"I don't think I know the card—"

"Yellow card, I.B.M., X-ray" he said. "Move it up."

I said "I don't think I have that one, could you—ah. Could I check on that one? Sir?"

"Move" he said, pulled the envelope from my hand, rammed his hand through its mouth, slapped the envelope against my chest, pounded a yellow card on top of the barrel, pushed me at a platform, pulled a lever and, when the barrel clicked, called "Chin on the rest, chest on the lens" and froze, waiting.

I pushed my face at the top of the machine, then closed my eyes and said "I got it" and moved away at a darkened door, leaving behind a clack and hoisting, bellow, and a fluoroscope of my hand crushing a cylinder of envelope, manila, brown.

And though I was not color-blind and saw a seventy-nine in green on a yellow ground, I could not see the giant E on the vision chart, and thanked the sailor slipping slides through the projector in the lightless room, said "Sir" when he told me out another door. And then I was not waxy of the ears nor swollen of the throat, and, when the doctor in the green cubicle turned his flashlight off and moved my lower jaw to close my mouth, I told him "Thank you, sir." I lost my group and held my envelope and swung my sock from door to door. I saw underpants grayed with filth and pus, rags held on by tensionless elastic, saw a tall boy, eyes as proud as a big bird's, march the floors from verdict to verdict in a union suit of white wool.

There were too many voices for hearing, then, too many sounds. I went at doors and doctors with a glaze of dark-green wallpaint layered on my eyes and the horse-radish smell of bodies jumping in my sinuses. The floor got hot, or my feet grew sore, I grew concerned for a poor score in blood pressure, went where a pointing finger said.

And then, my form reddened, the numbers of my waiting circled by the white-sleeved hands, I moved from silence to the hiss of boiling and a smell of something very very old. I came around a corner to a high-walled room with a stone-topped table and a trough, a handsome soldier, and a line of smiling boys.

The trough was hissing, the soldier shouting high and keen "You remove a bottle from that pile over there and you take it to that sink over there and you piss. Yeah, it's very ha-ha, it's very funny, but you better piss and piss fast because I am not gonna sit here all day and wait for you to work one up. Get the fucking *piss* up. *Move*."

I carried a bottle to the trough and squared my shoulders, listened to the spigots in the trough hiss, closed my eyes, stopped breathing, then began again and felt warm. And then I stared ahead and heard the crooning of Rogovin: "Oh you humper Jesus Christ come on come on come on for piss-off sakes for hump-off rubbing blue-wick oh come on come on come on." He crouched at the wide and crowded sink, his head aimed toward the high ceiling, fat knees pushing at the trough's rough rim: "Come on come on come on come on come on."

I carried my hot bottle to the stone desk and watched the handsome soldier dip a strip of paper in, drop it on a towel, and look away. "You wash that bottle out clean" he said. "You stack it neatly when it's clean and then you take *off*."

I washed and stacked, then walked back to bulbous Rogovin, fountain gargoyle waiting but gone dry. I said, "Good-by, Rogovin, I'm going home. Good-by. Go win the war." The trough hissed, the soldier called, the lubricated patients piddled.

He said "Come on come on come on come on" and I walked along the line, sock swinging, to the doorway of the last room. ("Come on come on come on.") I squatted and wrenched and stopped and knelt while the tape recorder chanted and a doctor watched and then I walked like a soldier, marching, swinging my arms and pounding my heels, to the barred window and Duffy and a room of silent new boys who waited to get naked and weighed.

I walked down marble steps that twined around the elevator falling *ratchet,* rising *ratchet,* and walked to Broadway with a U.S.A.F.E.E.S. meal ticket in my hand. I carried the green card to Broadway and Fulton, past newspaper stands and large amputated men on leather pads and men in gabardine and leather stores, down Fulton, then, to where the guts of television lay in bleached crates for casual picking-through by men who wore no ties, no gabardine, and then down stairs again, away from a hint of Hudson behind the leaning piers, to cold concrete and trembling long tracks, women buying chewing gum from bulbed machines, the slot I dropped an Army token into, and the peeled wood stile I turned to get to the platform, catch the train that bumped me back, by stop, by stop, to where a girlfriend yawned and lit the coffee up and asked me if I had to go away.

I told her no. I told her maybe. By the rung wide bed I told her *ratchet,* going *ratchet,* that I marked my time by her. And now, still marking time, I tell myself the high cuffs and pointed shoes of Rogovin, his fat not thick enough on marble steps in Grand Central Station as he pants down graceless to the long floor and cold tunnels, trains that go to warm counties where the beds are single, silent, numbered, set in lines.

I think again tonight of Rogovin, and the appetite I fed while water hissed and smells of boiling stood as stiff as Rogovin who, swollen by his own poisons, held by them, fooled by the mysteries of tissue, called "Come on come on come on come on come on."

I say "Good night. Rogovin: so long. Go win the war." I tell my noisy bed how I am here and how, in winter, with my sock aloft, I struck, rode home beneath the ground.

THE TROUBLE WITH BEING FOOD
1966

I was a very fat boy and always had to tolerate mezzanines in clothing stores called Big Guys and Muscle Builders, and in smaller shops in our neighborhood I would suffer comments from little men and women, spoken at my parents from between my legs, such as "He needs a lot of room in the seat, huh?" Then in college I grew thin without trying, and loved it, and wore as little as I could to show as much of my smallness as was possible. When I left school I ballooned again, and as I've wandered I've swelled. But Katherine, whom I travel to in Montpelier, Vermont, where she lives with her kids, from Cicero, New York, where I live with myself and the usual upstate ad agency—no talent in the shop, and little income—says she loves my stomach, which stays round when I lie down. She holds it sometimes between her hands. I try to cram it all inside her cold palms. I'm not in good health. I try not to pant on the pillow after love.

On the pillow after love at night in Vermont I hear my

heart knock, and it wakens me. Katherine snores. It's the same wet growl as my father's, which I heard twenty years ago or more, in our house at night. It's the same dream: the long thin lighthouse with its corrugated metal steps, and up the steps and around and up, *rumpty-dump* and *rumpty-dump,* here comes Casper the Friendly Ghost, footless but marching, *rumpty-dump*, and I waken to hear my father snore. I didn't know for twenty years that my pillow is a drum, I hear my heart. It haunts me out of sleep.

Katherine stops snoring and says "What is it?"

I say "Me."

She says "Oh." Then: "I thought it was one of the kids." Then: "Or Marlon Brando." Then she snores.

I say it to myself: Tomorrow morning I'm going on a diet. I'm losing seventy-five pounds. I'll become superb. Because when I have the heart attack I don't want my nurses making jokes about me.

I fold the pillow so it hurts the back of my neck, and I lie against my rock, a holy man, impressed that I'm not afraid, but earnest about staying up all night so as not to hear my heart do what it does in the darkness: surge to the base of my throat and rap like fists, race my pulse up, cover my forehead and neck with sweat. I am no longer impressed, and I *am* afraid, and I wonder if Katherine will waken to find my body in bed in her home, but no one home in the skin.

This is not a fertile pursuit. I consider her sons—slender like her, like their long-gone crazy father whom we often discuss, calmly and matter-of-factly, because Katherine and I are adults and this is her history: what can we do but discuss? (We can burn his clothing, cauterize her cervix of his trace, defile his name in the children's ears, and hire assassins to hunt him into terror and slow death.) But we discuss, it's what she needs. And in her old farmhouse surrounded by potato fields, wind with the smell of snow lying up against our breathing, I lie against Katherine, blink against sleep and the dreams of my fat body, consider her sons.

The question is whether Sears and Roebuck will question

my lie that Randy and Bob are my sons too. They're listed on the application I returned, which came to tell her about life insurance for less than seven cents a day, which everyone needs because in America there's death by accident every five minutes. It said "Think what a check for $100,000 can mean to your loved ones at such a time." All right: a fertile pursuit.

What I'm waiting for, of course, is the burst of pain up my neck, the tingling fingers. What I'm waiting for is a way to fall beneath a truck before that happens: accidental death, and an end, by the way, to night time snacks and the sneaking of seventh helpings—the ultimate diet. As Randy said: "Hey Harry? I think Santa Claus is fatter'n you, isn't he?" Not by much, kid. And when I show up in your fireplace you'll see who knows about ashes.

So here I am now, insured but still breathing, though not awfully well, at Katherine's living room window. The coffee is made, the house is in its early Saturday morning ease. Upstairs Bob rolls against the bars of his crib and they rattle, but everyone sleeps. The light swings through the town. It squeaks over fog frozen onto cornstalks that flap, and over the telephone wires fencing in the leafless trees on either side of the road. Everything had just been blue, and then it was ashen with cold sun on houses and fields. And now it's morning, the truck is idling at the trees beside Purdy's Bridge while its cherry picker hoists a man on its platform to prune the branches which in winter might fall under snow loads and snap the cable into silence.

While one man from the telephone company uses his chain saw and hooks, another in an orange safety vest gathers fallen branches and throws them into the back of the truck. Then he waits for the hoist to come down, then gets inside and drives to the next stand of trees, gets out and places the yellow sign in its metal frame on the road in front of the truck. It says

MEN WORKING

IN TREES

and he gathers more gray wood as the saw tears. I think of men in the crotches of all the trees on the town's main street, repairing shoes, restringing guitars, mitering wood, filing down ignition points. All of them are loved by fine women, everyone is smiling, and chamber music makes the shape of a room above the road and fills it. Yellow light from the top of the cab, in its squeaky swinging bubble, jumps through the town. And here comes Katherine, softly through the cold morning in her wooden house while children breathe upstairs and flatter us with their serenity. By keeping silent we pretend to give them cause for calm sleep: that lie of family love.

Think what a check for $100,000 can mean to your etc.

Heart disease makes you look *in*. So as Katherine walks across her living room—tall in a fleecy blue bathrobe that ties beneath the breasts and makes her look pregnant, big of foot in slippers of fleece that make her slide: long-faced, shining, glad—I hear my heart rock wetly in my chest. The pulse feels fast, I want to clock it, but I smile. She watches my eyes, she feels me sliding in and hooks me out as the light of the truck creaks by: "Good morning, good morning. Are you leaving us for good?"

I shake my head and smile. Good child.

She says "Are you leaving us for a quickie back home?"

Shake.

"Are you tired of older women? Am I scary-looking in the morning?"

Shake again. Reach for her furry front and pet it.

"So why are you sneaking around the house? You make coffee at dawn like a husband. Pad-pad-pad in your bare feet. Clank the pot like a cymbalist."

"That's me. Your community orchestra. Music to get laid by."

She pushes into my palm, says "Shut up," and we hug in until our crotches dock through cloth. We spill coffee, chunk the mugs onto the white-painted window sill, back off and circle around the sofa which is at right angles to the window: she goes her way around, and I go mine, and we meet at adjacent cushions. When I sit, my stomach presses up inside my body and squeezes my lungs. It feels like that. I pant, looking at the framed prints, the brown pots, cherry wood and clear-grained maple, cloth that wants to be touched. The hope, I guess, is that she'll look where I do, instead of at me. Or do I want her to watch me, and say *What's wrong?* so I can be brave, and start a fight in defense of not complaining—and thus complain while chalking credits up for courage, strength, great pain?

She looks at the walls and I grunt up onto my feet—a lesser stegosaurus in glasses and corduroy pants—and then I walk around, breathing. I bring our coffee from the window sill and hand a cup over her shoulder. When she bends to drink, I bow to graze on her long neck. She puts her cup on the mahogany table, but I can't reach there and, bending as I am to chew, I still can't set the cup on the floor. So I hang as if fastened by my teeth at great height. She feels this, then she feels the coffee droplets, then she turns—her face knocks my glasses from one ear—and when she sees my athlete's smile of teeth and flare of nose, she laughs so loud she wakes the children up.

Think what a check for etc. can mean.

So we go upstairs and get hugged by sleepy kids. Randy is talking already: he wonders if we can find an Indian long house or at least a war canoe buried in the backyard field which goes to the looping river. Bobby's trying to climb from his crib. Washcloths, turtleneck jerseys, miniature dungarees, small shoes, and all the time—"No, honey, put this hand through *here*"—as I help to dress someone else's children in a house he signed a mortgage for, there is a two-room apartment in Cicero, New York, where I am not listening to good opera on a bad record player while starting my survey of the week's new *TV Guide* to see what films I'll watch at 9 and then 11:30, 1:15.

Downstairs, Bobby watches Katherine fill a bowl with cereal and milk. He drops his spoon on the floor, smiles a sly one at me, bends to his bowl, saying "More?" Randy drinks orange juice and says "Mommy, I have dripping sinuses, I can't eat. Okay?"

Katherine says "No."

I say "Perhaps this isn't wise—"

Katherine, looking at my face, says "No to you too."

"I would like to marry this."

Randy says "I'll *up*chuck if I eat."

Katherine says "You better not, boy."

"Which one, Kath?"

"Both of you."

Which leads us to stacking the dishes, brushing our teeth —Bob chews a small brush ropy with ancient Crest—and the zipping of quilted jackets. Then, Katherine towing Bob in a wagon with wooden sides, we walk down the road toward the postal substation, Randy speculating on what happened to the Indians: "Then after the settlers shot their buffaloes, they got extinct. Like dinosaurs. They went into the ground."

"No," I tell him, "there are lots of Indians left."

"Uh-uh."

"Yeah, Randy. A lot of them live near Syracuse. A lot live everywhere."

"Then where's their spears and bow-and-arrows?"

"They're like us now, hon. They wear the same kind of clothes and work in offices—"

"Do you have any in your office, Harry?"

"Oh sure. Chiefs, too. Chiefs all over the place."

"Do they got any knives?"

"Knives? Listen—"

Katherine says "Let's be quiet for a while, Randy, okay? Let's listen to the morning for a while." Bob in his wagon is a motor, pukketing to the motion of his ride. Randy and I keep still. We hear woodpeckers and the snarl of jays, local dogs, cars on distant roads that are aimed for the Saturday errands I crave: the lumber yard drive, the haul to the local dump, the station wagon mission with kids in the back and no hurry, and then home to soup and soda and the wind blowing back from the river to gather about the house.

At the post office, which is someone's garage, Katherine and Randy go in for the mail while I stand with my legs apart and, holding Bob's wrists, swing him below me and back and forth. He shouts and laughs his wicked laugh, he's a lump of heavy cloth and knitted cap and scarf, his breath is small white smoke puffs. Then I put him back in the wagon—"More?" he says, holding his hands up, "more?"—and I listen to the knock in my chest, the brain pan's steam-whistle noises. I work at my breathing. I have eaten no breakfast and promise to starve all day. I breathe.

Randy stoops at the post office door and plays with a cat. The cat doesn't want to play, but Randy nails it down with his hand, crushing the soft neck to the ground, cooing "Ah, ba-by." Katherine comes out with some magazines and an opened letter. Her eyes are like the eyes in a drawing: almost like life, but too flat. She sends Randy ahead with the wagon, weaving in the road toward home, Bob an impossible engine.

"Dell's coming" she says. She shakes the letter out, it crackles and refolds. "He'll be here tonight or tomorrow. He wants to see the kids. Sure he does." We walk back between the two rows of houses on the town's main street, which is a road that runs in the country between larger towns. We say nothing, and her face is nearly not familiar, like the palm of someone else's hand.

Which leads us to the long lunchtime—I eat three sandwiches—and then we carry the boys upstairs for early naps. Bob's resigned, Randy is angry and wants to dig up Indians. The sky is smoky with early snow, and through Randy's window I see the black field hands nod their heads and tighten up. The farmer comes from his truck and lays a row of brown burlap bags beside a quarter-mile furrow which the tractor has made. They come from Burlington to work for thirty-five cents a bag. Now the snow comes down, a fine fast grainy fall, and Katherine and I lie down on her very historical double bed and listen to her children bounce around as the tractor changes gears and returns from the river over the field toward the road, pulling earth and potatoes up.

We're dressed. She's under the quilt, waiting. I say "Kath? You think I ought to go home?"

She doesn't answer.

I'm still breathing heavily from climbing the stairs, and I know she's listening to that too. I fight the lungs, the heart in its damp wrappings. "Listen," I tell her, "you should decide about this. It'd be easier, wouldn't it?"

She says "Easier for Randy and Bob, I guess. Less embarrassing for me. Weaker."

"No. What weak? It's your *life*."

"By now I should be able to deal with him. And you're a fact now, Harry. I don't have to get married for you to be a fact."

"But you *could*."

"I don't want to be married any more, kid. Shut up."

"Katherine: *I* want to be married any more." She doesn't answer, the tractor roars, a field hand's voice comes up. When I look over, her eyes are closed. I think of her driving me to the bus stop, and then the ride to Burlington with travelers and their old suitcases, shopping bags, cigars, then the wait in the gray terminal, the longer ride to Albany, then Syracuse, in darkness, and the half-lighted Greyhound station, all the people there not knowing me or that I've left a New England farmhouse and a family and people grunting over food dug up from cold soil. I tell her "I don't want to go back."

"What?"

"Go back. Leave."

After a while she moves on the bed, says "Then that's the decision."

"What's *yours,* Katherine?"

"I don't want to make love."

This requires a delicate pause, partly for the sake of dignity, and also because the idea now utterly seems to call for the act. I say "Well I don't recall inviting you, ma'am. Thanks anyway."

"Don't be lousy."

"Don't be private."

"No?"

Fat but not stupid, I say "It *is* a fairly private problem."

"That's right."

"Fairly private."

"Harry, would you believe it, I know what you mean?"

"I know you do. I'm sorry."

"This is dumb, Harry."

"Come on, Katherine, I'm apologizing. For thinking you don't know what's going on. And for your having to live through a week end like this. For trying to get you to say it."

"You're always trying to get me to say it."

"Sometimes you do."

"Porky Pig, I always do. Kind of."

"Kind of."

"How's your chest?"

"Beautiful. Bigger than yours."

"It's pretty nice."

She turns over and with one hand unbuttons my shirt, puts her icy hand inside, draws her knees up and becomes a small girl falling into sleep. We lie like that, and I reach to the bedside table for something to read—an old *Newsweek* with a puzzled article on heroes of the Green Berets recently dead in South Vietnam. Think what a check for $100,000 can mean to your loved ones at such a time.

And then four o'clock in the far western corner of the field, the burlap sacks in their rows, the tractor cutting the porridge of snow—it still falls lightly—and the hands in their thin jackets or only shirts, pulling up potatoes with the curved-metal long-handled forks, making deep noises, talking sometimes. We are near the river, its rich cold smell comes through the dense little forest on its bank, and Randy with a shovel impossibly long for him is digging with total seriousness through snow and hard ground to find an Indian long house or a fallen warrior's skull. Bob is on my back in a nylon and aluminum baby carrier, solid and happy and still, swathed in woolen cap and long scarf. I pant as I move with him, he listens to my rhythms and pants to my time: he says it.

A short coal-gleamy field worker in an aqua-colored windbreaker stands, stretches his back, blows on his hands. He calls over "You got yourself a burden, now."

I nod, smile. I have what I want for a minute, and he knows that. I say "Not as bad as yours."

He shakes his head. He calls "You want some of these for the missus?"

"Do I look like I need potatoes?"

He laughs and shakes his head. Bob laughs too. Randy comes over with a small lump of limestone. "Harry, is this from the bones of someone?"

I say "Probably. But put your *mittens* on, will you? Aren't you cold?"

His lecturing face comes on as he ignores mere weather to say "See this mark over here? This is where the bullet from the settler's gun went in. Isn't it, Harry? Here. You hold this while I go back to find the bullet. It probably fell out when his brains got rotten."

The field worker drinks from a pint of something dark. The cracks on his hard hands are white. Wind comes across the river to blow him into motion again. The tractor rips slowly past and Bob says "Choo!"

Then the man who harvests potatoes nearby says "This is a bad-ass day for living. You give me some other day for that."

Down the row a heavier man who is drunker—he slips whenever he moves from his knees—says "Pick the day with care, son. They coming bad more often. I noticed that."

The short one says "Your cold black ass told you that, isn't that right."

The drunker one says "My cold black *life,* son."

The snow is thicker—Bob says "Rice!"—and Katherine's house moves farther away, diminishes. Randy pokes with the tall shovel. An old green truck with snow chains in the southeast corner, near the road, is loaded with filled brown sacks. Bob says "Rice!" and then pants to my rhythms. The tractor starts toward us.

Then far away, at the distant house, a small car is in the drive, a man beside it. I see Katherine on the back porch. The man comes up the steps, stands below her, and they talk. She points toward us in the field. The tractor comes closer, Bob in the backpack stirs to watch. The man raises his hand, drops it, walks past the swing set in the backyard, then past the swings which the wind has set drifting on their chains, then over the chewed land in a fog of blown snow toward Randy and Bob and me. Arms across her chest, Katherine watches.

I look at the field hands and their long lives and think of *TV Guide,* of Katherine at my skin. Randy digs for dead Indians, Bob sucks for air because I do, the tractor comes closer, its steel fork tears up food and huge stones and think what a check for $100,000 can mean.

But the people in the story include that baby tied to the fat man's back. Everyone stands still, including Dell at the edge of his former freehold. Then Randy drags his shovel toward the man who waits, not looking anyplace but down, and I lug Bob back too, walking in the path the shovel makes. The tractor is past, there has been no accident. And Katherine watches us all come home.

Now there are the usual backstage noises: clatter of stainless steel and crockery, the battle of the kids being fed, the utter politeness of conversation among adults who cannot imagine how to survive the hours flat ahead of them—stony field they have to somehow work. When the children offer a chance, they drop all over them like sudden snow. There is the sound of corks being pulled and the tops of beer cans exploding. Now: here are the grownups at the kitchen table (it's a litter of chicken death and vessels), and here are the sounds of Bob in his crib too early to sleep, and Randy upstairs playing Odetta's song about it's being good to be home.

There is one partly nibbled drumstick on my brown pottery plate, and the wreckage of some servings of salad. My wineglass is scalloped from recent pourings. Dell, who has removed his sport coat and tie and rolled up his sleeves, drinks ale from a can—he's stowed a case in the refrigerator. His ironed-in shirt creases are still firm, and in his Oxford cloth he looks like Katherine's date, warming up for the evening's abandon. I feel as if I look like me: an ocean of rumples. Katherine drinks more wine. Some of it has run onto her thick tan sweater, and her hair is up, and I consider how important it is that I lick the wine from her front.

Lean pale Dell, with his left eye bloodshot, his large hand wrinkling empty Red Cap cans, his legs jiggling up and down, a smile on his long face—I sneak my looks at him. He says "Harry, you didn't eat much." The host.

"Well."

"I *know* you tend to put away more than that."

"Well I've got big bones."

Katherine, now my mother or my aunt, says "He ate a lot of salad. Didn't you, Harry?"

"Yes, ma'am. A good deal of salad."

Two wall lamps light the big room and Dell inspects the shadows. He says "You forget how intimate the kitchen looks."

"*You* forget" Katherine says.

Odetta celebrates the freedom of the eagle.

Katherine says "I didn't mean to be rotten" and pours more wine for her and me. My stomach cheers for political triumph, since Dell is excluded by his ale. But he pours water from his goblet into mine and holds the glass out for Katherine to fill, and she does. She looks at my plate. I slump in the chair and stretch my legs for better breathing; it doesn't work, and I sit up straight.

"So I'm a success" Dell says. "I'm a dean of students. What do you think of that? I'm into administration and right guidance." He drinks ale. "I will deftly guide them through the thickets of life."

"And along the abyss, don't forget" I say.

"Absolutely. Abyss, and crumbling ledge. *And* gorse and hawthorne and virulent ivies. Never ignore the virulent ivies. You get really fucked over if you fail to keep the virulent ivies in mind. I've always found that to be true, haven't you, Harry?"

"It's a safe rule to live by, Dell."

Katherine pours us more wine, and Dell holds his goblet up for more too, though he hasn't drunk any. She says "So here we are. The extended family." This is consummately humorous, and we all laugh.

Odetta discusses peace in the morning sunrise.

She puts her glass down and pushes at the stem with one finger, which suggests that she's about to suggest something. She says "I wonder why you came here, Dell."

I say "I think I'll take a walk. I'm taking a walk."

As I get my coat from the wall hook, not looking at Katherine, Dell stands up and takes his long black tweed dean's overcoat down. Katherine says "No."

Dell says "But it's your answer—that's why I came. I wanted to address the gentleman currently in your life."

"And see the children, of course" she says.

He says "Of course."

She says "Let's all stay inside."

But he is pushing my arm at the door and we go, not drunk enough yet, but going, and then already down the back steps and into the snow in our street shoes which fill with slush, walking past the swings and onto the field. There's a shape out there I wonder about, and a bright white moon, strong wind.

"Dell, don't you think someone should keep Katherine company?"

He strolls a little ahead of me, says "Why, someone always does, you see."

Now even though he's a dean, he's a dangerous man. He has beaten Katherine with his hands and once with a rolled-up newspaper they were using to train a Dalmatian which was killed by an electrician's truck. Of course, she's beaten him too —he's a dean. But Dell is drunk in a gaseous loose-jointed way that thin men have which frightens me. And he hates the history of their house. And he has to hate me too—unless he thrives by dining on pain which his liver, by habit, can turn into strength. He grips the cloth of my sleeve as we walk toward the river and—the moon turns it on like a lamp now—the stubby chipped station wagon snuggled into hard mud.

Dell says "I don't think my wife hates me any more, do you?"

"Me? No. No, I don't think so, Dell."

"Did you ever get divorced, Harry?"

"No, I never got married, actually."

"So you couldn't of gotten divorced, then."

"Right."

"Yeah. You're pretty young, still. So you don't know what me and Katherine are talking about."

"Well—"

"Unless you think fucking and playing house's the same thing as what Katherine and me're talking about."

"Look, Dell. This is very embarrassing."

"It *is?* Oh, I'm sorry there, Harry. It was not my intention to drive all the way here at risk to life and limb just to throw shadows on your soul."

"Dell, you want us to go back inside and have some more to drink, maybe? I don't know what to *say* to you. Maybe if we all got very drunk I would find it easier."

"Actually, old Harry, I am fairly well drunk at the present moment. And I don't honestly give two pounds of llama shit what makes anything in the whole world easier on you, kid." He lets my sleeve loose so he can indicate the whole world. "You got shadows on your soul because I'm a long-term cuckold on account of you. You put the shadows on your own soul, Harry."

By now I have stopped, and he stops too, near the light-colored station wagon. Around us the wide white field spins out, and the furrowed potatoes, the unfilled bags, curved forks. I decide not to discuss the logistics and amenities of divorce, or the question of when precisely a man is allowed to need the presence of someone without being digested by the major figures of her history. I do consider the gleaming points of potato forks, and Dell's deep craze, and how much a check for

$100,000 can mean at a time like this. Does homicide count as an accident if you really don't want to die? My chest is shaking at my clothes, breathing is serious business again.

Dell walks closer, he stands before me, takes my glasses off and puts them in his back pocket. "Being in the academic trade," he says, "I appreciate what these could mean to you." His breath is camphor and old cheese.

I have watched too many TV shows about the immorality of unpremeditated violence to be unwary. I am on my wet cold toes, moving backward, squinting at her blurry husband. And when he moves in again I scream a judo karate jiujitsu noise to paralyze his reflexes, I spin on my left foot and kick backwards with my right for the nerve complex just below his sternum. I strike nothing, something collapses in my ankle, I go down. He cries "You terrific bag of weakness, you don't snort the scraps off my plate!" And his knees or elbows land on my chest, my face is opening up beneath his hands. I push up, strike up, swinging wide loose powdery punches, get lucky, and something slicker than mucus streaks on my hand. He shouts —no words—and I stick up fingers as if I were a maddened typist. He screams, and then his breath is in close, his teeth on my cheekbone, he bites down and though I roll and kick and punch on his skull he bites in harder. I scream in his ear, I want to tear it with my teeth but can't.

He's off. His spit and our blood run along my cheek. I'm helped to my feet by people I can't see. I stand on one leg and hold to someone's hard shoulder. There's a smell of deep cold and blended whisky, sweat. Dell sits before me on the field, blurred face. I see the tail gate of the station wagon open—courtesy light, I remind myself. There are brown unfocused faces in the light, and much commentary.

Dell says "Like it's an academic situation, brother, dig? Much as I appreciate your interest, I don't think you see the subtleties here."

A deep voice near him says "I don't believe we your brother, *man*."

The potato picker I hold to says "You own one chewed

face, you know that, mister? I don't wonder if you got yourself some rabies."

The one with the deep voice far away says "Yeah, well that's the trouble with being food, son."

I listen to my body breathe and I whisper "Are my glasses broken?"

"If not, they the only things that's whole now. So come on to your home."

We slide and lurch to the house I can't see, me thanking and he saying never-you-mind, and both of us laughing once and then, by the time we reach the drifting swings, gasping in the cold air, silent. In the window above the back porch, there's a dim brown light. I say "Who is that? Upstairs?"

"A small kind of Indian. Red and yellow feathers. Watching you drag home."

"I wish he wasn't."

"Yeah."

"Hey—thank you."

"Uh-huh."

"Really."

"Yeah, all right."

The clicking of the storm door, and Katherine's face—fury? fear? no pain—and alcohol on the eaten face, an elastic bandage on the ankle, Randy's wagging headdress, the hobble upstairs, the weight of blankets, Katherine's insistence on silence, sleep, the sound of Dell's car starting down the drive: they wash into morning, the gray and golden early light in her still house, the curl of her body on the bed. Rather than consider, I twist down.

Rather than consider that an accident—by civil law, papal bull, Torah, or the New York Builders' Code—is what you don't make plans for. Rather than consider that the final sentence of the Sears and Roebuck contract no doubt says *In the event that the Insured is counting on this Policy for a measure of design in his little story, the Contract is nullified—it becomes just one more Cold-Assed Petition to end whatever pickle, puzzle, plot or unofficial war Insured can't deal with.* Rather than consider, truly, whether I heard Dell whine away. Rather than consider that I first heard Dell and Katherine yowl and sigh, make a long silence, and maybe leathery love, before he rode for home with part of my face in his war bag. Rather than consider shadows on my soul, or the thickets and abysses and the crumbling ledge.

My cheek feels hard and swollen, the ankle complains, but I twist down, slowly diving, and nose beneath the covers for her flesh. I push at the nightgown, kiss her cool thigh and crotch then stomach as she stirs. I come up onto her chest and suck a smooth nipple, turn at it.

She slams me, under the covers, and I sit up, the quilt like a shawl all over me. "Goddammit! Will you stop biting me?"

I wait, and my insides surge. Because here it comes.

She pushes her nightgown partly down, but not enough to pretend that we aren't naked, haven't been. And then she covers her eyes with her hand, whispers "I don't want to live with anyone, Harry. Not even week ends for a while. All right? I think just alone right now. All right?"

Her bare thighs and pubic hair and stomach sadden me, like spurious juke box mourning songs. I cover her with the blanket and put a pillow over my lap and belly-mound, I hold it with a hugging arm, she squeezes her eyes with her fingers, we wait. Bob bangs his crib slats to start the day.

It will not be a fertile pursuit. It will finish with a near sighted ride to Syracuse, the bullet-whipped fragment of a Mohican's skull another truth and trophy wrapped in my clothes. I will finish with an elevated leg, some great living stack of sausages and eggs and chocolate milk, and lean men easily breathing on the TV screen who smile. I'll use the extra glasses, an old prescription, nearly strong enough, which I've kept in a drawer with my socks for emergencies.

Katherine says "All right?"

Sure.

GROWTH
1969

In the waiting room of the surgical floor I stuff my head with patience and wonder what color a tumor is, I talk to myself about growth. Restless on plastic furniture, smoking a lot, squinting through the smoke at our time together, at the yellow light and dirty walls and silent doors that keep me from my wife whose body they are opening up right now, I open up.

By the time another match is in the filthy ash tray—candy boxes, butts, and broken matchsticks: bone-mounds of foundered fathers and children and wives—I am back and so is Anna, we're newly married, living in Greenwich Village on Morton Street in a large bathroom with cooking privileges. The knowledge we have of birth control pills is limited to speculations in *The Times* about cancer. Anna's mother died of cancer, and so we walk on Seventh Avenue, west to the Sanger Clinic, to make our marriage official and safe.

Anna wears her off-white tailored suit—she was married in it—and her face is very pale. I wear a sport coat and tie, and I think of buying Trojans over the drugstore counter in whispers. I think how old we're becoming, I cradle her elbow across the streets and feel us grow.

The clinic is in an old brownstone that's brightly lighted, the foyer is lined with posters about decision and self-determination and planning and how to make your life come out the way you think it should. A winding staircase to the right goes toward TESTING, and Anna whispers "Pregnancy tests—we don't go there now." She snorts as if she's told a joke, and I think of infinite rows of rabbits falling over in convulsions to announce a child is born.

We go to the left and into what was a living room long ago—birch logs stacked in a white brick fireplace, expecting no match—and Anna goes to the receptionist's window in the wall while I sit down, pick up *Time*, put it down, look at the green rug, look at Margaret Sanger's framed embattled face, look for Anna and see her go through a door, away. And this is the first time I actually notice the four women—each has sat as far away from the others as she can—who sit with their legs squeezed together, each one holding a large brown paper bag. Each takes her turn in staring—glaring—at me, and I look at *Time* again, then look at the rug, which still is green. I hear their breaths and sighs, the slippery sounds of skin against cloth, the whispers of their uniform bags. I think of the rabbits upstairs, falling over. I think of me falling over. Anna comes back, and she holds a brown paper bag. Her face is red and sweaty, and as she sits beside me on the vinyl sofa—it hisses and the ladies turn, then turn away—she whispers "Harry—my underwear."

I whisper back "What's wrong with it?"
"Bag."
"Huh?"
"My underwear—it's in the bag. They make you take your panties off."
"Oh God."
"Right."
"Everyone—"

"Right."
"Oh God."
"I think you're resented in here."
"I think maybe."
"Maybe more than maybe. You want to wait outside?"
"I can't."
"Why not?"
"I don't know. Yes I do: I'm scared to move."
"Well what should we *do?*"
"Pretend I'm your sister."

So we sit, the ladies pinging me with sniper fire from their eyes, Anna giggling into her white gloves, neatly folded, I, still working on an understanding of the Rabbit Test, looking at *Time* ("New Condominiums: A Place in the Sun"), at Margaret Sanger, at the unlit fireplace, the still-green rug. One by one the ladies are called, and others come in to resent me. I think of cold chairs against the naked skin and, when it is Anna's turn, I am nailed to my seat on the sofa by my erection, which I cache beneath a tent of glistening condominiums. Upstairs the rabbits stagger.

And then Anna, redder, her upper lip wet, beckons from another door. I shake my head. She says—aloud!—"She wants to *see* us, honey."

The ladies look. I say "Oh." I say "Well what the hell." I casually droop my *Time* before me and scutter to the door. Before I'm quite inside the office I say "Fascinating article here, really" and then I'm inside, next to Anna, on a gray steel chair in the darkened room, and the woman with dark gray hair is saying that she can't always *do* this but we *are* young and *just* barely married, and, well, maybe she can help us start off *right*.

She is Mrs. Nusbaum, with an office large enough for her, her desk, and her walls; she has three visitors' chairs inside, wedged and screwed so that anyone sitting anyplace can bump against everyplace else. She smells like strong mints and the perfume counter at Woolworth's, and her sibilant *s* sprays her breath: the closets of people long dead. The red plastic rims of her glasses are shaped like cat's eyes, and her teeth are perfect and white, her large hands luminous and lovely, strong dancing shapes in the darkened room.

She says "I've been counseling newlyweds for twenty-five years. *Real*ly. And I always say to them first off, what I say to those who, ah, *are*n't married: *think* of each other. Keep each other in mind.

"Now: Anna. How many times a week does intercourse occur?"

Anna grows larger, steamier, turns her head to me.

I say, definitively, "Eleven."

Mrs. Nusbaum says "Ah."

And Anna says "Five?"

I say "It varies. It depends."

Then there is silence, and Mrs. Nusbaum says "Perfectly *norm*al, of course. All I need to know is *rough*ly what the intercourse frequency is. That's fine, it's *fine*."

Anna says "I mean, sometimes people don't do it at all for a week."

Mrs. Nusbaum says "Married how long?"

Anna says "Two weeks."

"Perfectly normal, of course. It's often that way."

I say "It all depends, doesn't it? How can you measure these things?"

Mrs. Nusbaum says "Well, I'm not."

I say "It sounded like you were."

She says "This *is*n't part of the service, you know. I thought you *want*ed advice. Certainly, *I* don't mean to intrude."

And I, with my talent for dealing with human subtleties, tell her "Well I guess it just sounded like intending."

Her lovely hands settle onto the desk and she looks at them. Then she turns to Anna and says "Take care of each other, my dear. Just do what feels *right*. That will be all."

Anna, who is always begging to please whoever talks at her from the other side of a desk, says "We will" as if we've been blessed by a bishop. "We will. Thank you so much. Good-by, good-by."

And when we are outside again, walking back toward Sheridan Square, and I have caviled and groused, characterized Mrs. Nusbaum by her age and odor and diction, her office furniture, the hang of her hair, Anna says "I think she was pretty nice, Harry."

"Yeah. For a detective. For an investigator for the House Un-American Activities Committee. For the Kinsey Report."

"Scaredy-cat."

"Sure."

"Well *I* think she was trying to really *help*."

"Jesus, Anna, you're even talking like her!"

"I think she just wanted us to love each other and respect each other's body."

"How many times a week?"

"Is that fair? Is that really what she was like? You know, she wasn't telling us we didn't screw enough."

"You think I'm worried about that?"

"Are you?"

"You think that's what was wrong?"

"Well?"

"Shut up."

"You're such a big baby bear, Harry. I love you so much —let's not screw for a month and then have a marathon. I don't care."

"Shut up."

"You want to go over to Strand and buy a book?"

"Do we—do it too little?"

"We do it just right."

"Oh shut up."

We stand in the street and everything else keeps moving. She puts her hands on my face and says "Let's go around and buy things and I love you."

I tell her "You know, I got scared. In whatshername's office."

"You got embarrassed?"

"Scared. I still am."

"Oh bear—*why?*"

101

I watch the streets move past: "Because it's the first time I knew you were different from me."

Another burned match in the hospital ash tray. Discreet reminders on the public address system, hush of soft-soled shoes, claps of china from the coffee bar. The yellow light on worn vinyl. Faces of the other waiters, their clenched blankness. The fear and boredom, settled onto everything—on us, who watch for news to be born—like humid mist, like choking damp, like night.

Finding our apartment means turning onto Morton Street from Seventh Avenue South, walking to the small space between two high brownstone houses. The little wrought-iron gate opens into a long dark narrow alley, cobblestones worn smooth, which widens behind the brownstones and becomes a wiggly circle, grass and earth, flagstones, three enormous trees, a high wooden fence, and two wooden three-story houses where Aaron Burr's servants are said to have lived (this our janitor tells us, who is five feet tall, a woman in jeans with short-cropped hair who makes big paintings and fixes our plumbing several times a month). We are in the first house, on the first floor—there's one apartment to a landing—and when the door opens in, an eighth of the apartment is obscured.

The floor planks are wide and uneven, worn to a black tone of brown. The walls and ceiling are white clean plaster, the two windows onto the flagstone and trees are small and askew. At the opposite end from the door is our vast brick fireplace, charred as the bowl of a pipe, in which we burn scraps we find in the streets and fragments of our janitor's stretchers, since in the Village twelve small logs are five dollars the bundle. The bathroom, off the kitchen, is grand and white and holds our closet and storage shelves as well as the bathtub built for two. The kitchen is off in the corner of the room between the fireplace and bath: trough with spigots; Toledo Efficiency-Eeze composed of refrigerator box from the floor to waist height, four burners atop it, oven box above them to the height of the head; and, from the wall near the fireplace, at right angles, creating the kitchen "wall" on the other side of which, before the fire, is our dining table and its chairs, the two bookcases, waist high, which we've painted black and where we keep our Spam and Ann Page wax beans, our Woolworth's pots, our cannisters made of coffee cans: our stores.

The bookcases come from my childhood room, the table from Anna's aunt's garage; the stuffed sagging easy chair—to the right of the door, near the floor-to-ceiling bookshelves that lean out from their wall—is from Anna's first apartment in Pennsylvania, before that, from her aunt's garage; the camp trunk, which serves us as coffee table, used to hold my Camp Nok-A-Mixon uniforms; the bed is Anna's cousin's, the red plaid blanket we use as its spread was laid across the foot of my bed when I was eight. The books are mostly recent, and our tins of food, the clothes in the bathroom, the diaphragm hidden in the bureau, the bureau itself—from Sears—and the toothbrushes and soap. We are new also, and we grow in a garden of relics.

Because we're poor, I order a phone we can't afford. I tell the lady at the phone company business office that we'll need an extralong cord, since that costs more. And so we have a twelve-foot cord, which makes it possible not only to talk on the phone while sitting in the easy chair, but also to stand at the bathroom door and talk, or at the stove, or in front of the fire. Since I'm out of work, I spend our money, and Anna comes home from teaching—pale, eyes ringed black with exhaustion, shabby in an old ugly coat she bought when she was in college—to see me pointing proudly at the phone.

I put coffee on and demonstrate while she sits on the bed. "And you can move around while you talk, you know?"

"Gee, that's lucky."

"It's three dollars more, what the hell."

"You can put it on the floor outside the bathroom and take a bath and if you talk *loud* enough—"

"Listen, Anna. I can call up and tell them we don't want it. I don't care."

"No, sweetie, we can have it."

"Thank you."

"Well what would you like me to say?"

"How about something along the lines of why don't I get a job and *make* some money instead of spending it? How about did I look for a job today?"

"You don't look like you looked."

"I didn't even buy the papers."

"So: tomorrow."

"Stop looking so goddam brave, Anna."

She lies down on her back, still wearing the imitation tweed blue coat. She closes her eyes and I look at the phone. The coffee, of course, boils over.

I say "Death."

The tears come from under her lids like spring water squeezed from stony earth. She says "I'm not crying, don't worry."

"No."

"I'm sorry."

"Will you shut up with sorry, Anna? I'm sorry."

It sounds like a cough and sneeze at once, and she rolls over, curls herself up, holds her face, and shakes. The room smells of scorched coffee, and I turn off the burner and look at my boy's-room bookcases. She says "I'm sorry" and I hiss, and she cries hard.

"Anna: we're all right. We're all right."

She say "I don't *feel* all right."

"Are you sick?"

"No. You are."

"Anna, it's only a phone. Or my temper, whatever the hell it is. Only that. Don't make a *thing* out of it."

"No," she says, "I mean I think you're sick of being married. Living like this. Taking care of—"

"Yeah. Some taking care of. Letting you work your ass off, and I make phone calls on my extralong cord."

"You'll get work. It's a hard field to get into. Something'll come—I'm really not that worried."

"A little worried?"

"No. Sometimes. A little."

"Yeah, well you shouldn't have to worry at all."

"And neither should you. I don't want you working at some lousy job just for money. Something good'll come and we have to wait. We can *wait*."

I sit on the bed and say to her back "It's hard to wait."

"It'll be harder not to."

"It's hard to be married, isn't it?"

She nods her head.

I say "It was pretty hard *not* to be married."

"I'll take this" she says.

"I think you're crazy."

"Thank you."

"Well it's not a vacation."

"I don't think it's supposed to be."

"No."

"No," she says, "but it's all right."

"Really?"

"Yes."

"Really?"

Her hand is wandering, crawling, making moves. "*That*'s hard" she says.

I pretend I'm looking away.

"Our very own extralong cord."

I say—always saying: always—"I don't want us evading the issue, Anna."

"This is the issue. Shut up and listen. I think *this* is what we're supposed to be learning about. Making long-distance calls."

And we sleep on top of the bed, wrapped in the old red blanket, the old blue coat, to awaken in darkness, hearing from the rooms above us the voices of men.

A deep one—it burrs in the buried wall wood—says "I don't hate him."

A higher one, carefully controlled, says "I thought you were supposed to love him."

"And what would you know about that?"

"What I see. What I hear."

"So?"

"What kind of *so* is that?"

"It means so why are you coming around? So why are you here? So what do you want?"

"What you do."

"I guess you know?"

"I guess."

"Well you don't."

"Weren't you going to show me?"

The deep one says "You know—you're so campy, it's disgusting."

"How am I supposed to be? Confident? Is that how you'd like me? Well I'm not. I'm not. I'm not. I just—"

"You just want to have an affair, you just want me in your pants, you just want a slow piece of ass, don't you bullshit me. I know what you want. And you don't know anything."

There is silence, then a scraping across the wood upstairs, then Anna sitting up and pulling the coat and blanket higher across her chest, reaching for a cigarette, her hair swinging thick across her face like a heavy curtain. She says "Are they—"

I say "Yes'm, yes they are. *Boy*. Yes they are."

The higher voice says "I can go home."

The deep one says "If you could go home, you little fag, you wouldn't have come here."

The higher one speaks and we strain away from each other, reach with our senses, but can't hear. We stay apart.

And the deep one says "Fag, you cocksucking fag, you're such a *queer*. You know nothing. Nothing. Listen: do you know what it's like to fuck a hundred-and-eighty-five-pound man? Do you? *I* do. Plenty. You little fag queen vamp queer fag."

The higher one mumbles, and they both move away, to the far side of the room above. In the darkness I thumb my cigarette lighter and Anna bends over like a swan, arching toward my hands, not touching. The flame jumps blue in the darkness—white walls luminescent, cold; our old-time furniture like darkened ice—and it runs up the outer layers of her hair, across her brows and forehead. She shrieks and the voices bellow upstairs and whine. I push the blanket onto her head and rub, wrap, smother at her, shouting. The room stinks of chicken feathers burned. I rub at her face, and the voices bellow upstairs, I push, I push at her shoulders, tear the blanket away, I rub at her flesh, crush too hard against her breasts and she cries again—pain alone? just pleasure?—and she goes over backward and while they clatter and keen upstairs I throw myself onto her, into her, up, then I stop.

She is moving too, against me in the way that is with me, welcoming, and she still moves when I cease, then says "What? What?"

I whisper "I don't know. I got frightened. I don't know."

She says "No, come on. First aid, baby. Yes."

I say "I'm so *self*ish."

And she stops, as I have done, and we listen to the noise along the walls and through our ceiling. She touches my leg and we lie apart, uncovered, and we wait. We do wait.

And all that day those years ago, living in darkest suburban New York, before we moved back to the city, I waited. At the office I wait where I type in my steel and formica cubicle. There are four of us in the small writers' room and we all type on aged manual machines, and the bright blue carpet doesn't muffle much. It sounds like an engine room and in summer it smells the way it sounds. We each work on a counter at one end of which is a two-drawer file cabinet painted brightest of blues, and we all have blue phones, and we all have no doors, and all get paid rather little, and we're all too young to be here for life, and we're all without very much talent and it looks as if we'll all be here for life. We type out stories about education. Sometimes we go to visit schools, and sometimes we talk to teachers, but usually we stay in Greenwich, Connecticut, a few minutes from the New York State border where Anna and I live, and we write about making children's minds get born.

There are problems. Should the act of moving kids from

place to place in long orange vehicles be spelled out as *busing* or *bussing?* Are sliding room separators called *expandable* walls or *expansible* walls? Do we favor enamel sinks, which are porous and collect bacteria, and which are produced by a client who buys two ad pages every other month? Or do we lean toward stainless steel, which looks cleaner, except that scouring powder scratches it into porousness and it then collects bacteria, and which is produced by a client who buys *three* pages every other month. How do we talk about sex in the schools? And what about the public relations hack whose job is endangered by his total lack of ability and who telephones one of us once a week at least, trying to place his story about unwed mothers in the schools? What do we tell him to keep him from weeping us into embarrassment? And how could we accept and shape his story, which mentions his client's mobile health office, which he thinks can be converted for gynecological programs—"Thus, the district can spread more thighs with fewer stirrups at lower cost . . ."—at least in every other paragraph? There are problems.

Such as getting through some years of days there. Such as resisting the impulse to stand at the edge of one's cubicle, cry to Bernie the writer and David the writer and our Irish managing editor, whose flunked Jesuit zealotry stalks the sins of our copy, some shrill parody of house style such as "Small caps flush left boldface Does your district have a vandalism problem questionmark period paragraph The Pinwheel South Dakota School District did period paragraph And they solved it in a way that comma while it doesn't itals guarantee roman results comma might nevertheless be of help to you dash dash if you recognize these symptoms in your own bailiwick colon doublespace flush left boldface bullet Excessive broken windows flush left boldface bullet Greater dash than dash usual breakage flush left boldface bullet Rising frequency of custodial raids into your contingency budget period paragraph Here's what Pinwheel did comma and here's how you can steal a march on public school vandals period" and such as not staring at the salt sweat stain on Ellen's sleeveless blouse as she distributes the afternoon mail, and such as not calling home to see if Anna is back.

There are problems about the commuters at the Greenwich station—I am coming to look like them: we wear the same clothes and read the same newspapers, and I hardly can mind any more. If I had a longer ride, I would drink with them in the bar car and forget that I am forgetting that a job, we once decided, was a way of staying alive so as to live the right way.

And there are problems about walking from the Harrison station over the long empty field of high yellow weed and stunning sudden flowers, through an early summer hum of insect and chirring of birds, the smell of gasoline from Harrison's streets somehow comfortable among odors of dog turd and thistle and distant cut grass—the problems of Anna, who has gone to a doctor in Rye after school, and who might now be home at half past five, and to whom I will have to say my fear.

The banged-up car is there, in front of our landlady's house, and I go in and upstairs to our apartment—the living room and bedroom and bathroom up three steps, the kitchen and my study on the lower landing. I work in the kitchen more than in the study. Anna has made coffee and I smell meat cooking and I stand at the steps which divide our apartment, setting the empty attaché case on the floor, waiting.

Anna calls from the bathroom "Hey Harry? Are you here?"

"No, lady, this is the police. Your husband was raped by a secretary and abducted to Rio de Janeiro."

"Sounds good. I'm glad he's getting it *some*where."

"Hey, up yours, huh?"

"Can't take it?"

"Jesus, Anna, you ought to know."

Which of course means that no one says anything, and I leave the attaché case and go to the kitchen for beer. Anna comes in, and she looks white in her white terrycloth bathrobe. She kisses me and drinks a swallow of beer and gives it back. She sits at the table that once was on Morton Street, and once in Anna's aunt's garage. I drink some beer, and she lights a cigarette and coughs, cries out—up "It was so bad, Harry. It was so embarrassing!"

I stay where I am and close my eyes. When I open them she is blowing her nose and smoking and watching for me to ask. I hold the brown bottle at my lips and say around it "Did he find anything? Is he any good?"

"He kept acting like it was my fault I have an ovary that hurts. He looked at my temperature charts the way I'd look at some dumb kid's test. He was dirty, his wrists looked dirty—"

"Oh for God's sake."

"Well that's the way they *looked!*"

My voice gives back a tiny echo in the bottle as I say "But did he *find* anything? I don't care, you know, if he looked like Adolf Hitler. Did he tell you what was wrong?"

"He cauterized my cervix."

I think of soldering irons on the moistened skin. I think of smoke going up and Anna all white, humbled into silence and sorry for her pain, spread before him and being dutiful, scared.

I put the bottle down on the counter beside the refrigerator and say "What does that have to do with ovaries?"

"He checked them too. He checked everything. I have to go in Monday for a barium enema." She's crying again, and I hold onto the bottle as if the room is loose.

"For ovaries?"

"For cancer of anything. I don't know. All I know is I have to go in and get raped by some nurse with an enema bag. I don't know."

"Well. You're all right. I mean, he doesn't think there's anything really *wrong,* does he?"

"I do."

"Great."

"Well I do."

"Well you don't know."

"It doesn't feel like it's right. It hurts too much on my side. When I have my period sometimes it feels like something's bursting. So it doesn't feel right, and I think it's cancer, something awful, I wish you could have come—I'm sorry! I'm sorry, Harry. I really didn't want to say something like that because—of you—"

"What the hell is that *because of you* supposed to mean?"

"Because you couldn't come and I knew you'd feel guilty. It's your hobby, feeling guilty, and this isn't something for you to feel guilty about is what I mean."

"What you mean is because I was *scared* to come."

She walks past me to the refrigerator, gets us each a bottle of beer, then walks in her bare feet back to the table we have carried for three years in trucks, on top of cars, up stairs too narrow and down stairs too dark, and she says "That just isn't necessary."

"Hey Anna, you know why you piss me off so much?"

She smiles.

"Because you can get hysterical and then five seconds later you can *know* more than me. All the goddam time, you do that to me."

She smiles. She says "If I forgive you and you forgive me—"

"If I forgive me is the neat one, and I don't want to act like I'm singing this on TV or anything, Anna, but I spent an incredibly shitty day today because I didn't go with you. I should have gone and I don't want to discuss that any more. But I should have."

"Come with me next week."

"Okay. I think I can come."

Her expression is smooth, unseamed.

I say "I will. *I will.*"

We drink our beer and there is no mood left of anything, we are waiting for things to form in the air invisibly as if we waited for water to freeze while we watched. Anna turns off the oven, and we drink some more beer, and then, lighting a cigarette, rubbing cold water into her eyes, going back to the cigarette whose smoke wobbles up from the ashtray we bought at Pottery of All Nations, she says "He thinks it isn't my ovaries. That's why he wants me to have the barium thing. He says he *thinks* the ovary's all right. He should know: he had his arm up there to the elbow. But he doesn't think it's an ovary problem, unless it's polyps or something."

I think of great red bubbles of mucous growing against the bulbous shapes I imagine Anna to carry inside. "That's good?"

"I don't know. What it *does* mean is that maybe I can have children. He says after two years of trying, and if I'm all right—"

"It's me."

"There's a way of telling."

"Thank you. I've heard about it."

"He says we should sterilize a bottle and you should, ah—"

"Jerk off into a bottle, I believe is the terminology."

"And we take it to this clinic in White Plains, he gave me the address, and they can tell him about your motility."

"My motility."

"Honey, I didn't invent this business."

"So we can tell if it's my fault. No, I know. It's all right, we're not talking about *fault*. I know. Just motility."

Anna says "So if you want to—"

"Do I want to? Do I want to get a piece of ass off a Mason jar? Huh? Me? How often does a guy get a piece of ass that's a piece of glass? Me? *Sure* I want to. Yum."

Anna says "Well I have an idea about making it fun—"

And I walk out of the kitchen, across the dark little hall, into my study that is shaggy with books and the feathered pages of quarterlies. I say back "No fun, thank you. I don't think it's supposed to be fun. I'm going to do it because it's the thing I'm going to do. But no fun." I cannot stay with all that full and empty paper and I turn around and walk back in and stand behind her chair. I put my hands on her bony shoulders and rub at them, wondering which of the rooms has frightened me more. I say "Wait and see. All right? Let's just wait."

And in the morning I call the magazine and say I'm sick.

I call the school for Anna—she fears the phone receptionist's wrath—and then we boil a Spanish olive jar; Anna stays in the bedroom, brushing at the skirt of her wedding suit, which she'll wear to the clinic, and in the kitchen I watch through steam as the long jar rises and rolls, is rolled. Anna calls from the bedroom "Do you want me to help, honey?" Then, wickedly, but also with a shame (I decide): "Are you sure you don't want me to *do* anything?"

Holding the bottle with kitchen forceps, wearing only boxer shorts so as not to stain my trouser front, I call back "Yeah. I want you to stay in there."

"Are you nervous?"

"Of course I'm nervous. What if the bottle doesn't go down?"

Which is about what happens. Beside the aqua shower curtain and amid the smells of mildew, I sit with my shorts down like a sneaky boy, and try to test my motility. My erection has more to do with pudding than tissue and blood. My mouth is open, my eyes squinted shut. The forceps lie agape on the aqua bathroom rug. The bottle by now is cold.

Anna scratches at the door and I whine "Anna, get out of here?"

"I could lick it up and down."

"Anna. I can't sit here whacking off while you talk like that. It's perverted or something."

"So let me stop talking and start sucking."

"No."

"Harry."

"No."

"Please?"

"No."

"Harry."

"No."

And of course that is the rhythm I haven't had, the *Yes* and *No* of lovemaking, of writing in a stale study, of being a husband, of staying alive, and while my wife stands outside my door imploring, I masturbate my questionable sperm through the mouth of a sterilized jar. The fit is perfect, I hear my leather heart ticking moistly in my throat, I come like a time bomb which someone has set. The explosion is shameful, I sweat out of wretchedness nearly as much as from effort. While I cap the jar, Anna is calling, but I've used her as much as I can. I don't listen. I stare at my shorts and my ankles and wait to subside. I am locked from her, I'm my own secret lover, I am hoping to be discovered. With my sperm in my hand, I wait to talk and lift the latch and be found out, forgiven.

I passed the test, we were told to hope and try. Another match, and we are waiting. She, cold in the cold bright operating room. I, in the pale hot hospital hallway of chairs and ash trays. They are cutting her open and taking out a nodule of crazy flesh. While she sleeps—while we are waiting—a nurse carries down a frozen slice of Anna's meat, and in a microscope some stranger we might have laughed at years ago on Manhattan streets will take a look. He will decide if the surgeon upstairs should sigh and call the gynecologist, waiting in the hospital somewhere else, and tell him that the one-centimeter tumor over her heart is malignant, come upstairs to assist him in a radical removal of the left breast: slicing away the little droopy gland-sack and its tender nipple, and the flesh beneath the arm and on the back where the lymph glands could be cancerous too. And then coming out to tell me. Waiting for her in the Recovery Room to waken and reach up in pain, as if in scalding water, to see if she's still there.

This is what we've been waiting for: after those fast years, the slowest two weeks of our lives together. Our lives. From the GYN exam to now: waiting. With more difficulty, less conversation, every day. Waiting while we eat and while we sleep. Anna, the last few days no longer waiting with me. Holding herself away as if she wrapped herself in her hands and clutched herself like a sacred fragile thing. As if she were pregnant, we couldn't make love. Or barely touch. As if we both were waiting for birth. Which she has been—the birth of loss, of what could be delivered: her maddened tissue, new emptiness. Waiting while she listened and didn't hear, my words falling onto her as if from a distance, like fat cold separate raindrops. She, screened from any sense and emotion not her own: encysted. I, wanting to shout as if she were deaf. She, listening as if she were blinded and in a country whose language she never had known.

Does it all come out to this? The waiting, growing, fumbling through our bodies toward ourselves? The surgeon slogging in his bright green clothes as if he waded through slush —after all that growing? all that work? that clumsiness, error and faith?—to say "Okay. Now take it easy. Listen carefully. Here's the way it is."

Frozen section of years together, sliced and examined, these words—memory formed as prayer—are how it was. And therefore what it is. Here is what we have grown to. Here is what we have found: the flesh tearing away, the language less than adequate, the madness of continuing. I strike a match, the smoke goes up, I watch the surgeon come. I will speak before he does. I will tell him we continue. Along the perspective lines of the corridor, starting where the angles meet, and trudging vertically *down* to me, the surgeon comes with our future in his mouth. I blow out old smoke. I will tell him our courage. I will speak. He says, before I see his face, "Okay. Now take it easy. Listen carefully. Here's the way it is." I listen and say nothing, I see Anna reach up to her emptiness, we always were dutiful, we always had such faith, we were only obeying orders.

HOW
THE INDIANS
COME HOME
1974

Pictures from the forties make him sad about his parents: those baggy wrinkled unsynthetic fabrics, the graininess of the photographs, the hair pushed away from the faces—everything is frail, too much exposed, vanishing. He told me that on his eighteenth birthday his mother left a package at his breakfast place for him to find in the morning. It was a shirt box containing some pictures of him and his parents after his father came home from the war. And there was a long bright patterned apron string; pinned to it was a note: *Now you're not tied to this*. Doubtless, he felt free.

Before I divorced him (and almost didn't), we lived in our early days in a one-room apartment on Morton Street in the Village. He sat there once from six at night until the next morning, not speaking. We had just been married and both of us were scared. I kept thinking about the woman with two kids, he'd had a long affair with her—as if he were just divorced, and I had arrived in time to be his therapy, diversion.

I stopped believing that, it wasn't true, but *then,* with no money, and no plans, and just us and the little room, I thought of some earth-mother witch in New England, all breasts and belly, eating on his mind. It wasn't true. He told me later that he'd sat all afternoon and night not speaking because he didn't know what to say about being married. And because he was thinking about his parents—how he'd never know them now.

That wasn't our honeymoon. We didn't have money for a honeymoon. We went uptown and saw a Jack Lemmon movie, then had an ice-cream soda at a nearby Schrafft's. He was pale and funny in his sport coat (it was too small, he was getting good and fat), and his collar was too tight. He leaned over to me and said "I don't think I'll ever be free. It's my mother. I don't think—"

"You mean you love her funny?" I rehearsed everything I'd read about Oedipus in one psychology course at the New School and ended up saying it again: "You love her, in a kind of funny way?"

He laughed and held my hand. His hand was sweaty. He said "Freudyanna. I want to *mar*ry you. I want *you.* But I keep feeling like I deserted her. Both of them. I feel bad."

I told him something useless (I could have said the identical words about the death of a beloved dog, or the loss of a favorite scarf), and we continued—from jobs to other jobs, from the Village apartment to another one, from one hospital to another, from wound to wound, from childlessness to child to child. We tried.

We found out that we were in the West at the same time, as tiny kids. My father was in San Diego at a naval base. He wrote home that he was "traveling well," which got through the censor to tell us that soon he'd be shipped out with the division he was working for. My mother took everyone's gas coupons and drove in a 1937 Dodge, nonstop, through smelly motor courts and snow storms and communities quarantined with flu scares, missing the cable which told her not to come, that he was being sent out immediately. The newspapers tried to warn her, the state police tried to stop her, but she went.

And he wasn't shipped out, and we lived in San Diego and ate oranges together for months. I don't remember it—except a blizzard somewhere, when the car wouldn't go, and my mother sat at the wheel, crying. His mother took him on a train out West to see his father in training camp, and as soon as they arrived, Harry got sick. She had to take him home. He remembers ruining his parents' reunion. Why?

He remembers himself in a little khaki uniform worn under a bathrobe, walking through the aisle of a troop train in little slippers, talking to soldiers. Or he remembers his mother telling him about himself. He doesn't know which. Because he told me once that all he could remember from the war is himself and his mother and another woman, crouching in an elevator during an air raid, being scared. But America wasn't bombed. He really doesn't know how much of his past he's been in.

He remembers lying on a canvas cot in their dining room in Brooklyn, the walls full of bookshelves, light shining on the cherry table, but the room very dark. He remembers eating giant pills which tasted horrible, and his mother watching him whenever he wakened. She told him she was going to make him a surprise out of folded newspaper—an army field cap, a paper boat, a crown: he can't remember which. He says—he said this in Schrafft's, I didn't know then if we'd last for a week—the doorbell rang and his mother, who had sat beside him, her hair piled on the back of her neck, folding, cutting, went to see who it was. He can't remember the emotion, but he does remember the speed and sureness of it, tearing at the cap or boat or crown, rending it, undoing it, wadding it to trash. He remembers that when she returned, she wept. He remembers that in the dark room, between their table and their books, she said "Oh, I'm sorry. I'm sorry." In Schrafft's, the night of the day we were married, he said "I shouldn't have done that. I wonder why I did that."

He remembers the dream he had when he was very small. It was after his father came home from the service, and their house was filled with dreams. He would see, in his sleep, the flesh-colored blanket folded at the foot of his father's bed. It was almost flat and smooth, but there was one wrinkle in it. He would see the wrinkle and be sickened with terror. He would see the wrinkle and whimper to himself in catastrophe. Then he'd get up and walk very quickly across the hall to his parents' room, stand at the head of his father's bed—never, he remembers, looking at the blanket there—and touch his father's forehead, saying "Daddy, Daddy, I had a nightmare." His father would waken in fear; he remembers that whenever he was suddenly wakened his eyes looked frightened. Then he'd hold his arms up and pull him into bed and hold him for a minute. Then, holding his father's hand, he'd be led down the hall to the bathroom where they both would stand and urinate, then spit afterward into the toilet. This, his father explained, would make the dream go away. Then he'd be led to his own bedroom and would fall asleep, at peace. He remembers not wondering if his father went to sleep again.

After the separation and divorce he came in very frequently, sometimes every week, to see the boys and sit with me in our old apartment (I shifted the furniture so he wouldn't feel too sentimental), and talk about what we all "needed": money for clothes for the boys, a vacation trip, new drapes, whatever he could provide, so long as he received no benefits. The idea was for all of us to conspire that he hadn't ruined our lives. And he hadn't, we had all worked together on that, and the boys were surviving (I think they were relieved), so we had a couple of our pleasantest months since the early days. As he relaxed, he came in less frequently, and after a while we lived on a schedule instead of by our needs. Or the schedule *was* our need. It was something of a relief to finally be a typical couple, divorced.

This Tuesday he called, that Thursday the boys called him, on Saturday we met if he was in town, twice a month he made sure to get in from upstate, and once a month he took the boys for a long week end, picking them up in his car and driving to places I hadn't heard about, where they stayed up late and ate motel food and were happy together.

It was going all right. Ian often wet his bed, but he did it less and less, and I refused to believe that Harry and I had all that much to do with it. So few fifty-year-olds wet the bed, Harry always said, when we discussed it on the phone, and we agreed that he would grow away from it when he was older. Stuart had no problems. He ate incessantly, beat on his older brother when he felt like it, and asked for his Daddy when he woke. I hated that, and I hated his confusion, and my days always started wrong because of it. But you can't have what you want, and sometimes you live with wrong mornings. Ian cried at school sometimes. All right, you sometimes cry at school. I wondered when Harry cried. And I knew when I did.

One night, when we were married and not asking questions out loud, I patted Harry on the shoulder in bed and said "It's all right, it's all right, sweetheart." He'd been whimpering like a baby, the saddest long moans.

In the darkness he said "Thank God. Thank you, I couldn't have—" and he rolled over onto his side, then onto his back, said "Thank God you did that. Thank you." I rubbed his head, which was wet with sweat, and he started to fall asleep again. Then, suddenly, he said "We were in a hotel room, it was all white. You were dressed up, very elegant. You were dressed up for other people. Not me. And you kept walking around this bright room. You looked so *tall*. You kept saying you were leaving me. That it didn't make any difference—you really didn't care, Anna. You were completely independent."

"I'm not leaving you, Harry. It's all right. I'm staying right here."

"I felt abandoned. Totally abandoned. I mean, it was *grief*. I didn't know what I would ever do. Then you noticed —you pointed it out to me, like you were *appalled*—you told me I was urinating in the bed. I saw the top sheet going all dark. But I was so relieved. I felt so much easier. Then you woke me up." I stopped rubbing his head and lay back. "Oh, Jesus" he said. "Jesus, I was like a *baby!*"

"And I was like your mother? Abandoning you?"

"Jesus" he said. Then: "No. It was *you*."

"Yes, but it still could be—"

"No, Anna! You didn't *feel* like my mother. You felt like you."

"That's who I am" I said brilliantly.

And after a while he said "That's scary, if it wasn't about you and me. I mean, it's scary *enough* if it was you and me. Horrible enough. But if—"

"No. Never mind. You shouldn't do that with dreams. Just have them and get them over with and forget them."

"You don't think it was anything else?"

"No."

"Really?"

"I'm staying right here. I'm here." And soon he rolled over and reached across my shoulder into my nightgown, seizing, squeezing, holding on, and we made love.

Harry remembers, and I do too, the night, before we had Stuart, when Ian stood in the doorway of the kitchen, his pajamas too short, his feet so long and bony, face too white. He kept rubbing the corner of his mouth, his lips were so delicate and swollen then, his fingers so long, the nails all bitten away. He was holding the imitation army rifle made of metal and hardwood that Harry had bought him for Christmas. He held it by the barrel and stood like a sentry at the edge of the kitchen—*our* sentry, almost guarding us from us: because for a minute we stopped. Harry was in his plaid wool bathrobe, I remember, and I was wearing blue tights and a dark blue woolen shirt, half open, the buttons were torn because he'd been tearing at me. A family portrait. I turned away so Ian wouldn't see my eyes: "Hey, Ian, you're up so *late*."

Ian said "I heard noises." I guess he was looking at Harry, or me, I didn't watch. He said "Did you hear a shouty kind of noise?" After a second he said "I thought I heard kind of shouts."

I was washing my eyes with cold water. I remembered when Ian was supposed to be napping and Harry and I sneaked into the bathroom. I had his pants open, and my dungarees were down, when Ian pushed the door in and it caught on Harry's foot. He knocked and said "Hey, what are you guys *doing?*"

By the time Harry let him in, he was squatting on the floor, like an Indian, and nearly zipped. I was sitting in the tub, the water was running, I was splashing it onto my blouse as it rose above my half-closed jeans. Harry said "I'm watching Mommy take a bath. What're *you* doing?"

Ian said "Watching you watch, Daddy." And then, knowing something—his eyes got that sly look—but knowing nothing, he said "Aw, you guys are silly" and went to bed.

So, anyway, I washed my face that time in the kitchen and moved as fast as I could so he'd see as little as possible. I tucked him in and kissed him and told him something about a little argument, but don't worry, and then went back. While that was happening, Harry was standing in the kitchen and looking at the brown bean pots and white pottery vases and dark blue bowls. The copperware hanging on the brick above the stove caught a lot of light, and the beauty of the apartment, its investment, made him bow his head down. When I came in he was saying "The usual self-pity. And it's right on time."

I said "What": a dull heavy stroke—the weariness I really felt, and then the weariness I wore for effect, to show how much he'd exhausted me, and also to show how little his statements could mean.

He answered with the same weighted voice, because by then we were puppets on the fingers of our arguments. We were left with nothing to say. The arguments were having *us*. We must have looked like kids at the end of a street fight, panting and drooling, surrounded by cold-eyed watchers: us. He said "I was feeling sorry for myself. And hoping you'd hear me. So you'd feel sorry for me."

"That's fair."

"It wasn't supposed to be fair."

"No," I said, "I mean it *was* fair. That you told me. I have to have a cup of tea, my stomach hurts. Do you want some?"

So we sat in our little dining alcove, drinking sweet tea and not fighting. We were exhilarated by loss and fatigue and we were talking very quickly about the people next door, making jokes about the neighbors calling the police. Then we grew silent, shook our heads, rubbed at ourselves. Then we just sat. The sounds of sirens from the street came in, and the horns of fire trucks, the humming of the elevators going past our floor. I felt process going on, and us being sheltered away from it. It all felt very easy. We drank our sweet tea, got drowsy, were still.

I remember noticing myself, but not thinking anything. I stood up, and he got frightened. I took the tea mugs away to the kitchen. He said "I thought maybe we'd talk?"

I felt such confidence, it was like punching him: "Oh, we talked."

"No, I mean together. Come on—you know: not fighting."

I said "Well, we *don't* not fight, friend. Do we?"

Harry said "We've been doing a lot of it. We've been doing it. But I keep thinking"—he closed his eyes and lay his fists on the table—"it's so *temporary*. Really. You know? We get past this time, we work on it, we get back into the *marriage*. That's what we have to do, Anna."

"Long distance."

"What?"

"From away from each other. If we think, we can think in different rooms. Different cities, probably. I don't know. But different. Away."

"You're pulling that on me again?"

"I'm not pulling anything again."

"Oh no. Just the old Now I'm leaving, so you shape up—that's all."

"Harry, I'm not pulling."

"That's blackmail, Anna."

"No. It's moving out of the apartment."

"With Ian, of course."

"With Ian."

"And that's kidnaping. He's my boy too."

"Listen, Harry, you can call it murder if you want to, but in the morning Ian and I are taking a trip."

"You'd threaten me with this."

"No, I'm doing it. If you think it's threatening, maybe you should think about what makes me do it."

"God, Anna. Look: first of all, I don't want you to go. Second of all, if you're so dumb and you have to go because you want to *hurt* me. Uh, third, you can't take Ian."

"Second, third, eleventh, a million, who *cares*? You're such a bigmouth, Harry, you know that? Kidnaping, murder, blackmail—you make that little boy cry in his bed—"

"Don't you use that on me. You were there too, remember? You were there. Don't you tell me I made him cry."

"You did."

"You made him cry *too*. Don't *tell* me that."

"Don't you believe it, friend—that you can cry your way out of it this time. This time you pay. So you can cry *blood* if you want to and I don't care. Because you terrified that baby."

"We can't do this, Anna."

"Done. Consider it done."

"You're crying too."

"So *what?* Somebody's always crying here. So what? See, the idea is *not* to cry. The father makes the wife and child *not* cry. The wife makes the husband not cry. It's called being happy. We don't *do* that any more. So you and your words and your crying—oh, do break it, Harry. Break it. You can throw everything in the apartment at me and smash it and break it and everything and I *still*—"

I saw Ian's rifle come around the corner of the doorway as I held my hands in front of me and Harry crushed the round brown pot against the sink. I made a noise as if he'd kicked me. So he kicked me in the leg. I kicked him back. He hit me where the jaw goes into the neck. I swung back with both hands moving, and he caught my wrists and held them in the air. I screamed, I bit my lip, I cried and kicked at him. He dodged my feet and bent my arms back. I heard it over the noise: *Shh.*

Ian was in the doorway, making his playtime rifle sounds, holding his gun awkwardly because we didn't let him watch war movies yet. Elbows pointed out, and long feet, the white tight face with its giant eyes, his mouth rubbed red, pointing his rifle and shooting: *Shh. Shh. Shh.* We heard him.

We also heard Harry that night, later, when he decided to exhibit all of the characteristics for which I did not marry him, or love him, or find him more than tolerable. He pouted, always where I could see him (in an apartment, not difficult), and he looked like a sullen ape; instead of peering through banana trees, he sulked over sofa and chairs, from around the refrigerator door. He sighed very loudly and talked to himself, but always less than audibly, trying to tempt me to ask him "What?" and open the conversation again. I put Ian back to bed and calmed him down. He was eager to sleep and make his escape. I took a shower and stayed in the bathroom a long time. Then I went to bed and had stopped shaking enough to read our old magazines, one after the other, while Harry sat in the dark living room, going past the bedroom corridor from time to time for more ice. He was widening the purview of his exhibition to include The Drunken Husband in Grief. I read the old *Times* magazine supplements; struggling Africa and threatened Kurds fell through my head like garbage down the dumb-waiter.

I don't know what he drank, but he must have drunk a lot of it, and I must have fallen asleep. His voice from the bathroom, echoing, woke me up. I heard the water splash in the tub as if an elephant were taking a foot bath, great slopping wallops. Then it dripped on the tile floor, very loudly, and then he stood in the corridor, fully dressed and soaking wet and in his shirt and pants and shoes. He came in and turned the bedroom lights on, whistling, and I just watched. He tore his shirt buttons off, pried off his shoes without unlacing them, dropped his trousers onto the mound of wet clothes. When he was naked he said "That's a comfort," sighed very loudly, and lay on the bedroom rug.

I was thinking about rape, and making love, and how

cold he must be—his big nipples on his hairy fat breasts had shrunk and tightened—when Ian walked slowly into the room, pale and blinking, his pajama bottoms in his hands. A smear of moisture ran halfway up his tops, which he still wore.

He said "Did you wet yourself, Daddy?"

Harry said "Oh, hon—"

"Did you wet yourself too? I did." He wept, and his noise was uncontrolled, as if his body still were asleep. "I did too."

Harry, still naked, with a very frightened face, moved before I did, picked him up, carried him through the hallway to his room. I followed, listening to "I got you" and "I'm taking care of you, love" and "Don't worry, it's just an accident" and "Here I am, I'm here." He was tender and very worried, clumsy, not so awfully drunk-sounding then. I loved them both. And I didn't know how to tell them apart. I stood in the doorway as Harry crooned to him, taking off the pajama tops, folding him in a curve on the bed, rubbing his flanks, talking. He opened drawer after drawer in the half-darkness until he found underwear and fought him into the shorts and shirt. He tore the wet bottom sheet from the bed and nearly knocked Ian to the floor. Ian began to cry again. Everything fell out of Harry's hands, he had to do everything twice.

I whispered "Let me help."

He said "Stay there, Anna! I'm here! I'm *here!*"

He pushed the door shut and I leaned my shoulder against it to listen as he slammed the drawers of the blanket chest, saying "Okay, Ian, I'm taking care of you. Daddy's here, love, you're fine, you're fine." And Ian, mostly asleep, still cried.

After a while it was silent on the other side. Then Harry said "Hey, Ian."

"What, Daddy. What?"

"Remember that time you went for a walk with me in Prospect Park and I told you about—"

"Go way, Daddy. Go way."

He said "Ian?"

"What! Go *way!*"

I went into the living room and cried. There were bottle

and glass, moisture rings on the coffee table, the feeling that something sordid had happened on the cushions and carpet. He would follow me here, puzzled, not very drunk now, without an idea of what to do. He would follow me anyplace else. He would follow me into the living room and then I'd take him to bed. In the morning I wouldn't leave. I cried harder. And then he came in, still naked, tousled, bewildered, sorrowful with guilt for all the moments of the evening—mine, and Ian's, and his own. I looked at him and rubbed my face and said "Very nice."

He smiled his kid's grin and said "I haven't a thing to wear."

It was after Schrafft's, and during his first month at work, when he told me why his mother, whom he never saw, might be troubled. I hadn't asked. I was sitting on the bed, being exhausted (we had no couch) and he was doing the dishes in our only room. I said "Maybe you haven't done anything wrong."

"No, I probably have" he said.

"Maybe you *didn't* do anything wrong."

"Maybe." He stood with his hands in the water, amputated by the suds, and he sighed, his shoulders slumped, he shook his head. When I remember this he looks so young to me, much younger than a man in his thirties, inexperienced. He said "It's scary. It's like I'm always looking down two tunnels at once, two tubes. One has what *we* feel, how pissed off we get by all that—*that*. And down the other there's all this information, I don't mean that. Excuses. Or, I don't know—*possibilities*. You know: she doesn't feel well? Her kid's grown up and away? Her life isn't too fine for her any more? She's in trouble? Lonely? You know what I mean. I was standing here, looking down the tubes. I sort of moved my head to the left, I always see ours on the left. The one on the right is hers, right? It went dark. But that's not it. I *made* it go dark. I knew I was doing it. Just now. I just stopped looking there. I don't want to *know* that side any more. It's either us or them, so it's us. Finished."

"Harry, I never heard of anything like that. I don't believe it."

"No? *I* believe it. I said to myself Hey Harry, that's too easy. You just don't want to work at seeing both sides. You know what I answered myself, just now?"

"How could I know anything like that?"

"I told myself Okay. Call it easy. Be *bad* from now on. But all you see is *your* side, yours and Anna's. Otherwise you're dead. So now I'm not dead. But maybe they are. Maybe she is."

"I don't understand."

"No? Well why should you? Why should you? That's what happened, though."

"Does it make you feel bad?"

"I don't know. Really, I don't know. It makes me feel *good*. But maybe bad also. I don't know."

We're civilized, we see each other a lot now (or we're *un*civilized and see each other a lot, I don't know). He misses the boys, and me, and I miss him and hate that he's so sorry. And he's admitted to me that as tough as it is for him to be away from his sons, it's tougher to be with them now, though he's doing a pretty good job of being with and being without. We remember each other, and treat each other well. We're both still confused. In some ways he always was a child about sex—our first two years together were difficult for him, for me. He grew up, though, and I did too, and that part was all right. But I know that he remembers it, still is sad and ashamed for what we were in the beginning. And I'm certain that he wonders now about the men I know, and how I know them. We only discuss the past.

He remembers so many books, he always remembered them: *The Book of Fascinating Facts,* Red Randall on chipped yellow pages fighting the Japanese, Gregor Felsen's *Navy Diver,* with sheath knives cutting air lines for the sake of democratic life. He read them after the war, of course, when he lived in his imagination in his room. He memorized a book called *America's Fighting Planes,* which had full-page pictures of the B-24

Liberator dropping bright red bombs on German cities, and the Grumman F4F Wildcat pumping tracers through the canopy of a Japanese Zero and into its thrashing pilot's back. In the fifties he wanted to join the Civil Air Patrol as a ground spotter to help intercept a Communist sneak attack; he was disappointed for weeks when they told him he was too young. He remembers Zane Grey and Edgar Rice Burroughs in old torn editions he brought home from his Boy Scout library. He knows the plots and names of so many of them, and he hardly remembers his life.

When I think of what he's told me it's like watching an inferior print of a very old film—jumps, splices, long illogical gaps. He remembers, and yet he's mostly forgotten. A lot of him is lost. I asked him when his grandmother died and he didn't remember. I asked why she died and he didn't know. He does recall that he dressed for school on the day of her burial, and came downstairs and ate his breakfast and went to the door, while his father watched him and his mother watched them both. He didn't know about funerals, or doesn't remember knowing; or, if he did understand, he didn't realize that he was going to school on the wrong day. His father silently looked while he stood at the door. His mother said "Aren't you coming to Grandma's funeral?"

He doesn't remember speaking, but he knows he went to school. Twelve years later his mother told him he answered "What'd she ever do for me?"

I told him he'd never have said it. I said she shouldn't have told him, even if he had. I told him plenty more about her, and he said his recollections probably didn't do justice to his parents, he simply couldn't remember enough.

I said "Harry, what do you feel *good* about?"

He smiled a helpless little boy's smile and said "Just you."

I have never been so frightened since then—not even during my operation when they cut me apart, and not during the divorce when something similar happened. Not even now, when I sit on our furniture at night and think I could marry him again if I got crazy enough. Who's sane?

He remembers so much and so little. Well I remember too: the Christmas when Ian was so sick and we gave him phenobarbitol and penicillin and he always slept. We put him on the living room sofa and took turns watching him. He cried sometimes when he woke, and once he said "I have to make a teepee!"

Harry said "Bathroom, honey? You have to go to the bathroom?"

Ian pointed his finger down at the blue and white blanket: "The Indians don't have any place to *live!*"

So I fed him cream soda and Harry took the elevator down and went to gather twigs on the city streets. When he came back, Ian was sleeping, he looked flat beneath the covers, his white face on the white pillow was completely still. Harry sat on the rug and tied three wet crooked sticks together with twine, then looped the twine around the sticks at the base so the teepee would stand. Then he very carefully cut out shapes from the Sunday *Times* and taped them on as walls: Richard Nixon and Archbishop Makarios for the Indians to go through to get home.

Ian woke up crying. Harry rubbed at him, he kept pushing the hair away from Ian's forehead as if he wanted to see him better. Ian turned his face away from the hand and kept crying. Harry said "Look, baby. Hey: here's a house for the Indians to live in. Pretty good, huh?"

Ian looked at the little wood-and-paper tent and his face twisted in, his lips looked redder. He cried and shook his head, he said "It isn't good enough, Daddy."

Harry didn't know what to say. I don't know what he remembered, but he had counted on his medicine working and he'd failed; I thought that he'd cry too. He said "Ian, hey— tell me what's wrong and I'll fix it."

Ian said "You didn't make a *door*, Daddy. Why did you do it like that?" He tore at the teepee and the newspapers fell from the frame of twigs and twine.

PREPARE
AND
FOLLOW ME
1975

Sixteen years after he was graduated from college I telephoned the alumni office and asked for my son's most recent address. They gave me mine. I hung up without saying thanks. I called his hairy friend, the one who sells drugs to his students, and asked if he was there. He wasn't. The friend asked if Harry was in trouble. I said I wouldn't know, he was only my son. I hung up. There was a woman he used to see, he never mentioned her after a month or two of mentioning her a lot, but I remembered the town she lived in, and her first name. I sent a post card to her first name, in care of the post office in her town:

> If you happen to see my son, would you be so kind as to remind him about his parents? This is very urgent for us. Thank you,
> *Claire Miller*

I went to the offices of *The Village Voice* and fought my way through the homosexuals and lay psychoanalysts and paid for a classified advertisement:

> HARRY MILLER. Communicate with your parents right away. This is very urgent.

The New York Review of Books charged eighteen dollars for a small classified advertisement in their personal columns, among the retired school teachers seeking sadomasochistic sex, and *The Times*, for an advertisement among the notices of boats sailing away and husbands stopping their wives' charge accounts, wanted more. I paid. I sent a cable to the cottage he used to live in near his college. The Western Union boy would find his body if he were sick or dead. Then Mac did his duty as a supportive husband by telling me they telephoned telegrams. Nothing got delivered any more. But I sent it: URGENT YOU CALL US SOONEST REPEAT URGENT LOVE MOTHER AND DAD. Mac sat in the apartment with his bad heart and fatigue listening to me type out letters and make my calls. His own son! And all he did was listen and watch.

I asked him "Don't you agree with me?"

"That we miss him?"

"That we have to get in *touch* with him!"

He shrugged the way he always shrugs. Spread his hands out feebly like an old man. Smiled. Said "We may *want* to get in touch with him. But he's grown, he's officially a man. He's free to do what he wants."

"Free." That was all I could say. If he doesn't know, what I want to say won't tell him. Free. Nobody's free. If I were free I wouldn't humiliate myself this way. Free. He could be starving in a hotel someplace needing help. *We* could be. He'd never know it. We're old people. What if we needed assistance? Don't adults care for anyone but themselves? Don't they feel responsibility? Do grownups act like children? Isn't there something called consideration any more?

Free.

I wrote to his former teachers, in case he was communicating with them. The boy who used to be his roommate in college. That girl he lived with before he was graduated. She was the only one who wrote back:

> I'm sorry you're upset, Mrs. Miller. Harry and I haven't been in touch. I have a baby girl. I wonder what you have in mind. Why should Harry write to me any more?

Everything is *any more*! As if you stop caring about people. But they were only children when he thought he loved her. I can understand children not loving each other after a while. But do children stop loving their parents? That is an *any more* I do not understand. Intolerable. I've never read such a vapid letter in my life.

Even though I was old and therefore useless, the publishers continued to give me free-lance jobs. One of them was editing a book on colonial American funerary art. Tombstones. I was behind schedule, mostly because of the time I spent on finding my child. The author was pressing for early publication because he was afraid that someone else would publish the definitive work before he did. A young professor. Very ambitious, and always talking about art. I never understood whether he cared about the subject or his fame. He had written too much about the dead. He was too philosophical, and I was trying to prune his prose. He wanted everything to stay in the book. The letters he sent me! Crying out about integrity and the sacredness of tombstones. But really, I think, he was worrying about his reputation. And a book club selection.

The rubbings and photographs were fine. I was fitting his

clotted paragraphs to the illustrations, paring away the language, specifying a less dramatic type face than the one he'd seen in a book on Egyptian temples. I realized I had memorized some of the funeral verses chiseled over dead young wives and infants—

> As you are now, So once was I,
> In Health & Strength, tho here I lie,
> As I am now, So you must be,
> Prepare for Death & follow me.

Most of the inscriptions were variations of that kind of *memento mori*. It's what was placed beneath the clumsy death's-head and the stonecutter's signature. On every stone. Imagine signing a gravestone!

Someone called Beza Soule had signed the wandering stone which my professor wrote about so eloquently. Most of the stones stayed where they were. Holding the bodies in place. But my ambitious author had found one that traveled. He wrote his most lyrical prose about it. Dorcas Cooley had died in December of 1785 in Orange, New Jersey. Several years later Samuel Cooley, grieving husband, had needed to move. My author followed him. Through family Bibles, village histories, even bills of sale. He tracked him to upstate New York and then to Vermont. Where he went, Dorcas did. My author found them both in Port Henry. Wrote a disquisition on how tombstones offered permanent truth—time and place were fixed, there wasn't any guessing in graveyards. But here was a stone that traveled, death couldn't pin a grieving husband down. I liked that. It interfered with the flow of the book, but I kept it all in. Beza Soule had chiseled:

> From death's arrest no age is free
> Remember this a warning call,
> Prepare to follow after me.
> The wise, the sober and the brave,
> Must try the cold and silent grave.

Yes. So I was doing three jobs at once. Living with Mac on Riverside Drive while Columbia tried to force him into retirement. Finding Harry. Editing the sadness of other people, dead so long and past all mourning. They were data. It was excellent practice.

Then in the mail, while Mac was teaching, I had a short letter from the executive editor. Would I please send them all artwork, copy, and correspondence. They wanted to review the book's status. I thought I was going to throw up. I made a cup of tea and sat in the kitchen with the letter. I even reread the envelope. Review. Professionals don't talk that way to each other. Or if they do, both parties know what is meant. I did. I was losing the assignment. Review. Once they had the materials, they would send me a little note about a change of plans. My ambitious author? No. He had no influence. My ambitious executive editor? Probably. But it didn't matter. The letter had been sent and I had read it. The machine was moving and wouldn't stop. I wrapped everything in cut-up supermarket bags and taped the bundle heavily. I pasted an envelope on the outside. In it was my letter:

Dear Irving:

I thought your ethics were stronger than this. If I've been taking too long, it's because a family tragedy has been on my mind. But that's not important. You wanted the book ready for production and it's not. Perhaps you could have gotten in touch sooner if there was such a hurry. I know. You did call several times. But you really never warned me this could happen. I suppose it's terribly urgent, or you wouldn't abandon me like this. I do hope the project works out for you. You have my deepest wishes for a successful career. You're very good at your job.

Sincerely,

Claire

I put on my raincoat and went downstairs to mail the package. The streets were filled with degenerates and dog droppings. There were policemen everywhere, in cars and walking in pairs, and I thought there would be another riot. But it was just another day on upper Broadway. At the West End Café, where Mac's graduate students seemed to spend most of their time, several young people came out and pushed into me. The package fell and one of them, a girl with nipples showing through her tee shirt, stopped to pick it up for me. She smiled. I looked at her and she shook her head and said "Is something wrong? Are you all right?"

My eyes filled with tears. I shook my head and walked away. It was hot with a raincoat on, but I was wearing my nightgown under it and had to keep it buttoned. I thought I would faint from the heat. But more than anything, I wanted to send the package off and never hear of it again. I knew that I would look for its announcement in *Publishers Weekly*, though, and have cramps when I saw it. Mac would ask what was wrong. I would tell him. He would tell me not to worry. I was crying.

The post office was very cool and dark. I put the package on the green table and rested for a moment. I was shaking. I mopped my face with one of the cotton gloves I'd found in my pocket. Then I took a deep breath and after a short time in line I told the man at the window that I wished to send the package first-class and registered. He asked me for six dollars and pushed the levers on the postage meter. The ticket slid out and he glued it onto the package. He was a short black man who wore a coat and tie and smelled of lavender. He smiled and said "Six dollars, ma'am?"

I held onto the window ledge and closed my eyes. I took a deep breath, opened my eyes and said "I'm afraid I made a mistake. I neglected to bring my purse with me. I left in such a hurry, you see. My purse—"

The woman in line behind me sighed loudly. The man behind the window stopped smiling. He said "I already sold you the postage, ma'am."

"Yes, but I don't have my *purse* with me."

"I sold it to you, ma'am. What am I supposed to do about that? Pay for it? I don't make that much. I can't afford to treat the customers until I get a raise. Hey, come on, huh?"

"I don't have my purse. I'm sorry. I'm very sorry. I left in such a hurry, I forgot to bring my purse."

"La-dy."

"I'm sorry" I said. I tore the wrapping paper with my fingernails and handed him the piece of brown shopping bag wrapper with the postage on it. I kept saying "I'm sorry." I left the postage in his hand and carried the package away. An illustration fell out of the torn seam of the bundle and I got down on my hands and knees to pick it up. I put the picture of someone's cemetery marker under my arm and stood up. I bumped into people in line and said "I'm sorry." I went out into the street and ran home. I thought I would vomit from the running.

I had to stop several times on the stairs of our building because I was so short of breath. My ears felt clogged. I wondered if this was how Mac always felt, working against his sick heart. He never would tell me. Inside, I dropped the package on the floor and took my raincoat off and threw it on top of the package. I was crying out loud. I said words I don't say. I went to the telephone and dialed the last number Harry had given us. The same man answered and when I asked for my son he said "I told you last time—he moved. Hey, are you all right? He *moved*. He really doesn't live here. Really. You sure you're all right?"

I hung up. I unbuttoned the neck of the nightgown and went to the bathroom, ran the tub. I said "I am all right, thank you. I am fine. This is temporary. I'm fine." I didn't look at my fat old woman's body in the shaving mirror. I sat in the tub. I closed my eyes, cupped some water onto my face and leaned against the cold porcelain. I let my legs float up in the hot water. I slid down until my toes bumped the spigots. I let my arms float up. I lay on the water and ignored the fact that I was there. It was excellent practice.

Then it was as if I had just awakened. Nothing told me what to do. I made my body move. Walked across the bathroom, very weak. From the medicine chest I took one of Mac's razor blades and unwrapped it. I threw the wrapper in the wastebasket. Went back into the tub. It was like going back to bed. I ran more hot water until the steam made my face sweat. There was a whistling in my ears. It was from outside. I'd put tea water on to heat. Everything whistled. The razor blade was covered with moisture. It looked cold. I held my breath. I pushed the blade on my wrist. I put my hands under the water between my legs and pushed harder. I waited. I wasn't pushing hard enough. I tried. Both hands went down deeper together. When they rested on the bottom of the tub and when I was crouched over myself I pushed again. It didn't work. And when I pulled my hands from the water suddenly, there was a suction and a surge and splashing. I threw the razor blade at the sink and cried again.

I leaned on the edge of the tub for a long time before I got out. There was a small slice on my left wrist. A little accidental cut. I went to the medicine chest, dripping water on the floor. There was nothing to take that wouldn't hurt. There was iodine and there were several bottles of tablets which Mac used. I didn't know what they would do. There was Clorox behind the toilet on the floor. I opened it and smelled and gagged and put it back. There wasn't anything to do. I let the water drain from the tub and watched it go down.

Then I went into my bedroom and dressed. Soon I was ready. I went into the kitchen. There was no whistling. The kettle was empty by then. It was hot, there was a smell of scorching. I turned the stove off and sat. Then I went to the refrigerator and took down a tray of ice cubes. I put a bottle of Scotch and a glass on the table and made myself a drink without water. Eventually I used some of the water which had melted into the ice cube tray. I noticed nothing remarkable during that time. That was all I noticed—that I wasn't aware of the color of our kitchen, or the texture of our chairs, or the weight of our glasses. The stinging on my wrist didn't go away.

It was an open slice, the little flap of skin was loose. I had slanted too much. I should have torn in and down. The stinging stayed, and I told myself that our kitchen had blue vinyl wall covering and our chairs were brown wood with blue fabric and our glasses were heavy and cheap. I went into my workroom which used to be our dining room when we used to entertain and came back with a legal pad and a felt-tip pen. I made notes:

1. Buy more Scotch.
2. Rewrap package for Irving.
3. Pay $6 to man at post office.
4. Pick up razor blade from bathroom floor.
5.

I couldn't think of what to write. There was nothing else to do.

Mac came home from Columbia. I heard his footsteps and the pause outside the door. He was catching his breath so I wouldn't worry about his heart. Or so I wouldn't worry him. Then the lock tumbled and he came in. The delay while he put his briefcase down and hung his seersucker sport jacket in the hall closet. And probably hung my raincoat up as well. The delay while he took more deep breaths.

Then his footsteps in and a pause again as he saw me, described me to himself—the bottle and ice cubes, what my face must have been.

"Well" he said.

I said nothing.

He said "Hello."

I nodded. I felt my lips purse.

He said "Is everything all right?"

"Everyone keeps *asking* me that" I said. "It's a stupid question."

"Everything is not all right, I take it."

"Congratulations" I said.

"What happened?"

"Nothing. Nothing new." I started to pour myself another

drink, but that was simply too theatrical. I folded my hands on the table. I said "They took the job away from me."

"The graveyards?"

"Tombstones! I've told you that. It's tombstones, not graveyards."

He sat down and loosened his tie. I saw how sweaty his collar was. His face was very pale. The heat was bad for him. He nodded. "What happened to your hand?" he said. "Your wrist."

"I had an accident."

"What kind of accident?"

I laughed. I said "A very minor kind."

He took my hand and turned it, looked at the inside of the wrist. He stood up and went to the drainboard, saw that it was empty, opened the knife drawer. Closed it. Went into the bathroom. I heard him grunt and sigh as he bent down. I uncapped the pen and crossed out number 4, Pick up razor blade from bathroom floor.

He came back in and sat. He opened the bottle and poured a small drink for me, dropped in what was left of the ice cubes. He looked like a very old man. "For Irving Lehrer?" he said. "For *him?*"

I was disgusting. Mucus blowing from my nostrils onto my lips. My head shaking like a crazy child's. Hands trembling on the glass. All the noise I made. He stood to come and hug me but I shook my head harder and made the motions of pushing in the air. He sat down again. Put his handkerchief in my hand. I wiped my nose with the sleeve of my blouse, then rubbed at the sleeve with my other hand. Rubbed the hand on my skirt. I didn't raise my head.

He said "Not for the assignment. Everyone loses assignments."

"*I* don't lose assignments."

"That's not true any more" he said. "Now you're like the rest of us."

"I don't want to be."

"I'm afraid that's irrelevant now. You're the same as we are."

"No."

He didn't answer. I listened to him reach for breath. He was so old. And I was. I stood up and went to the sink. Washed my face and dried it with a paper towel. With another paper towel I rubbed at the slime on my blouse, my sleeve. I turned to lean my back against the sink and watch him watching me. In our blue vinyl kitchen with wood-and-blue chairs and cheap heavy glasses. I said "Will you tell me something, Mac?"

He nodded his head.

"Even if you don't want to?"

He looked at me, then nodded again.

"It's something you don't talk about."

"All right. I'll talk about it."

"Harry."

"All right."

"Do you miss him?"

"You say *I* ask stupid questions."

"I mean really *miss* him."

"Yes."

"I miss him so much, I think I hate him, Mac. Is that comprehensible to you? I hate my son because he's left us."

"Everyone leaves everyone, Claire. Children grow up."

"You sound like Dale Carnegie, Mac. For God's sake. Answer me *really.*"

"No I don't hate Harry our son because he's grown up."

"That's not what I asked. Because he's *abandoned* us! *That's* what I'm saying."

"Well. Abandoned. That's interpretation, isn't it? Abandoned. He's gone, is all. He can't find peace with us and peace is what he needs to find."

"But what about *us?*"

"We're all right."

"We aren't."

I walked over to where he sat. His face was taut, it was

younger-looking. It seemed less pale. His eyes were still too yellow, but he looked somehow better. He said "No. We aren't."

"And what do you feel about him?"

"Sad."

"More, Mac! Say *more!*"

His eyes became wet and his chin wobbled. He clenched his teeth and stuck his chin forward. Daring someone to hit him. "I'm—I'm not—" His chin wobbled again and I thought he wouldn't be able to speak.

I said "Please."

"I'm miserable, Claire." He was crying, then. "I'm very unhappy." He clenched his teeth harder and the words stopped. Then the crying stopped. He was the old man who slept in the bedroom down the hall from me. I nodded my head. He said "I sometimes think I've lost my life." He said it lightly through the trembling, and he tried to smile. He didn't smile.

I went closer to him and put my arms around his neck. I pulled his head to my stomach and squeezed him. I pushed him back, then, and bent. His eyes were closed. I bent closer and kissed him on his tight dry mouth. I forced his lips apart. In spite of his struggling I pushed my tongue between his lips and past his teeth and in.

THE GOAL OF LIFE IS DEATH 1975

When he woke up he was in a small dark room with a picture of Jesus Christ on one wall (holding a little baby on His lap in some arsenic-green garden where fruits and vegetables drooped from the trees), and on the other wall there was a picture of a doctor holding a huge hypodermic needle above an infant's bottom while young parents of the forties (it had to be Norman Rockwell) gasped and shielded their eyes. Orange leather furniture, textured wallpaper, shelves with heavy books, a desk with many papers weighted by a stethoscope. "So I didn't die" he said.

The woman with white stockings and a white dress, and bright red plastic high-heeled shoes, said "How do you know?" Her face was very pale, her hair very black, and her nose by itself was beautiful. On her face, with narrow eyes and a torn-out mouth, it was one more sign of distress.

He said "You mean I *did* die? Nah."

"About six mothers with little babies carried you in here. It looked like a war going on, you were tripping and throwing up—"

"I'm sorry."

"We got you clean."

"For all the—fuss. I was just going for a walk."

"Some walk. Babies crying, one of the mothers was crying, she got hysterical. A welfare mother. Doctor Hebner sent for an ambulance after he looked at you."

"I don't want an ambulance!"

"Tough titty. What's your name?"

"*Tough titty?* Do nurses talk like that?"

"Didn't you hear me?"

"Now that you mention it, yes, you said tough titty."

"You can hear all right. What's your name?"

"Miller."

"You have a cardiac history, Mister Miller?"

"Not much longer than Seventh Avenue."

"That's what he thought."

"Doctor Hebner?"

"Jungle Jim. I'll go get him. Lie there."

"Yes ma'am."

"Miss Poteri."

"Yes, Miss Poteri."

He sat up and rubbed his chest and shook his head and did the usual things. He'd repeated the same gestures for years. Once he went through his repertoire on the 116th Street station of the IRT when he dropped his brief case, bent down to recover it, stood up without it, leaned against a girder, and simply went away. He woke up, or recovered, or started breathing again, a few seconds later, leaning against the girder, spit all over his chin and his heart against the wall of his chest like a crazy squirrel, all claws and climbing. It was a matter of waiting each time. In the IRT he waited until he could bend over for his brief case and then did, and it was gone. A courteous kind of city: they stole the bag, but let him sleep. He rubbed his hands over his nose and mouth while Miss

Poteri let the door slam closed. The crack she'd gone through had let in sounds of infants weeping, dozens it sounded like, and women talking loudly, bright light, a feeling of heat. The crying of babies raged at the door but didn't quite come through.

Then a tall black man in a light blue linen coat came in and closed the door. He was bald and the color of coffee with a minimum of cream, and he shone as if he were wet. He wasn't, his hands, with long fingers, were dry and hard. He shone a light in Miller's eyes, made him open his mouth, told him "Quiet" in a soft voice while he timed the pulse. Without saying anything else, as if Miller were a child, he took his tie off, folded it on his sport coat which lay on a chair, helped him unbutton his shirt and take it off, then pulled his undershirt up to his nipples. He said "Shh" when he put the stethoscope onto his chest and bent in front of Miller's face. He smelled like medicine and aftershave lotion. He put the stethoscope on Miller's back, told him to hold his breath, then stepped away. "Trousers" he said.

"Excuse me?"

"I put my hands on my patients, Mister—"

"Miller."

"Yes. Trousers?"

So he undid his belt—Hebner gestured that was enough—and lay down on the sofa while Hebner squeezed at his stomach and abdomen and groin. He went to the end of the sofa, took Miller's shoes and socks off, ran his hard hands over the tops of Miller's feet. "You can dress" he said. "But stay on the couch. I was looking for signs of obstruction" he said. "The pedis dorsalis sometimes indicates if there's blockage. Also the femoral, in the groin. Don't be embarrassed."

"I'm not."

"No? Then you must do this a lot. You've had other incidents like this one?"

Miller nodded, said "I'm sorry for the disturbance."

"Quite all right, Mister Miller. I'm glad those ladies found you near here. Do you know what you were doing?"

"Having some coronary insufficiency?"

"Sitting in front of a delicatessen window on the sidewalk with your legs out in front of you, turning white and crying."

"That's new."

"The crying?"

"Also the delicatessen."

"I've sent for an ambulance, Mister Miller. I'd like you checked over by a cardiologist."

"I was. Yesterday."

"What'd he say?"

"I was making progress and two more vitamin shots and that'll be thirty-five dollars, please."

"A German Jew."

"Excuse me?"

"All the German Jews over sixty who still practice medicine in New York give vitamin shots to people like you."

"Oh."

"I don't think you're having an infarction" he said. "But I'd like you to ride the ambulance in and get looked over. Will you do that?"

The door opened in and a small person—like a miniature man of about five feet, sloping shoulders, very short arms, small hands—whose face looked like those faces you see in films during bank robberies said, through a cleft palate or muffled throat, "You gonna gimme a bloodshot, Doc?" His voice was a boy's, as far as Miller could tell, and his eyes were bright and alert. It was like looking through a stocking mask at a bandit: nose smeared fat and to one side, the mouth a droop and hitch at once, the face much wider than deep. "Hey, Doctah Hebnah, you gonna gimme a bloodshot? I *hate* dose bloodshots. You gonna gimme a bloodshot?" He was pale and happy and blubber all over in his jockey shorts and low white socks. He clasped his hands in front of him and stood.

"Hello, Matthew" Doctor Hebner said. "This is Mister Miller. He's sick."

"Hi, Mister, uh, uh—I forgot."

"Miller" the doctor said.

"Hi, Mister *Miller!*" he shouted. "I remembered! Very pleased to meet you, I'm sure."

Miller said hello and watched them. Hebner went to the boy and put a hand around the back of his neck. "Matthew's a borderline educable. He's fourteen and he can read at the fourth-grade level and learn at the third-grade level, and we're proud of him. He went to school all by himself this year and he's finishing in triumph and going to camp, right Matthew?"

"I like camp" Matthew said.

The door opened again and Miss Poteri came in, pulled Hebner's hand from the boy's neck and said "You're behind again." Then she said "Come on, Matthew, I want to weigh you."

Matthew said "Thanks! I needed that!" and went out the door which Hebner slammed behind him and Miss Poteri. "You know anyone who'd work from one to nine, get paid for working one to five, and not mouth off at me for seeing five patients an hour instead of twelve?"

Miller said "Yes, me. I just retired today. I'm out of work. I used to teach at Columbia."

Hebner nodded. "Maybe that explains it."

"Are you a psychologist too?" Miller tried to smile. Hebner walked out.

Miller put his shoes and socks on very slowly, with rests between each extended movement. Professor Miller, Person Emeritus, belly bulging from his pants in a pediatrician's office, waiting for an ambulance to carry him in triumph to a waiting room writhing with drunks and addicts and people who were dead. He thought of Hebner's hand on Matthew's neck and said "What could you possibly need Jesus for?" He thought of the little classroom, and the students in their chairs and on the windowsills, the scuffed wooden podium, the pitcher of chlorinated water on the table beside it, their open notebooks, their expectation of his voice. Miller's last lecture:

What ever became of the Vorticists, my wife inquired the other day. I told her I didn't know: our conversation for the morning, pause for laughter and, hearing none, proceed.

They are doubtless still spinning in their graves, pause and proceed. However, I told her, perhaps in our conversation of the evening, the Realists still live, though no one here has asked. Their flabby spirit resides in those who tell us what we know, and what the book reviewers want to know they know. They offer us information about hotels, airplanes, international banking, and popular psychological disorders.

Our next Realist blockbuster will be a novel called *Depression*. It will be made into a film, possibly before the "writing" is finished, called *Sad Claire,* which will be directed by Sam Peckinpah and which will feature a slow-motion fifteen-minute examination of cannibal corpse cunnilingus. Later it will become a Broadway musical named *Sad!* The woman who in the novel committed suicide here will be a lustful Latin American dancer named Tang who tours the German-crowded bistros of Honduras during the late fifties. She is not happy. So she dances, clad in a turtle shell, to the music of a two-hundred-piece orchestra, in order to seduce a former SS colonel whose spirits are low because no matter how cleverly he hides the Israeli secret police keep finding him. In the last scene, as a result of eating a special mushroom given her by a free-lance filmmaker who unrequitedly loves her, she has been transformed into a Third World revolutionary named Natasha. She knocks over a bank in order to provide her lover, the colonel, with enough money for his escape to Washington, where he has been offered a job with the State Department. He kills her—as she dies she sings "Is This What a Woman Is For?"—to preserve the secret of his identity. Her Transactional Analyst, a Gay Liberation wrestler and former Green Beret corporal, sings her lament: "Too Bad You're Getting Fucked Over by Existence."

A producer named Kafka makes eleven million dollars. A literary lawyer named Burdoo takes home two million. The writer, Eisenhower Skirt, buys two condominiums and a plane with which to commute between them. He leaves his wife—it is the domestic epic, my friends—and lives in both of his houses with a fifteen-year-old former rock group publicist named

Golden, about whose appetites, and his inability to satisfy them, he will write his next novel, after receiving an advance of two hundred and fifty thousand dollars. That "book" will be made into a movie called *The Loose, The Limp, and The Wobbly,* which will then be transformed into a matinee success named *Jello!* in which Helen Hayes will play the lovable and crafty campus sex researcher who is trapped by skyjackers in a Cherokee tank loaded with Arab terrorists. And that is what's become of Realism. If you've taken notes, kindly send some to my wife, to whom I neglected to say this or—pause for laughter—too much else this morning.

As a character in an unpublished novel will doubtless say one day (whispering on pages in a Doubleday mail room or a brief case clamped between Harcourt Brace Jovanovich knees commuting to Greenwich with the tenured Columbia professors and the media executives), we must prepare to have failed to say what matters. It is one of the risks.

You'd think they'd put gin in these pitchers, if only for an old man's final say. I will drink, and pause for laughter.

Hearing none, I will continue. It occurs to me that through these long years of professing what we've conspired to call English I have been troubled more by one problem than all others: if you wear light-colored slacks to teach in, and you go to the bathroom before your class convenes, you run an awful risk of emerging before your students with pee-stains on your pants. In my undergraduate days, the professor I most admired taught while standing before us for two hours with his fly buttons open. We prayed that he might never discover the depths of his exposure to us. No doubt he did, though. Still, a gaping hole is somehow less disgraceful than the smeary stain, I think. These are occupational hazards some of you will one day head for, should you muster the energy to continue writing small articles on minor aspects of miniature authors. Wipe yourself front as well as back is my advice. Don't be frightened of unorthodoxy.

Miss Poteri came in, looked at him, said "You all right?"

He nodded, reached for his shirt.

She said "I'm going to sit here a minute, all right?"

He nodded again.

She said "The goddamn ambulance is supposed to *be* here. Probably, some jig way uptown cut some whore's heart out—I mean, some lady who *deserves* her welfare money—and the doctors decided the meek should inherit the earth. You know what I mean?" She lit a cigarette and leaned back in Hebner's orange-covered swivel chair and sighed.

In 1920 Sigmund Freud completed a dense long poem called *Beyond the Pleasure Principle*. It was translated by C. J. M. Hubback from the second German edition and was published here in 1922 by Boni and Liveright, whose tradition of discovering and gambling on exciting new work—they did the early Hemingway and Faulkner, Jean Toomer's *Cane*—still deserves our admiration. This in itself is remarkable, since so few publishing houses deserve very much of anything. The bulk of what is published we can divide into two categories—ah, your notebooks and pens: you're preserving me—Labia Literature or Poshlust. The latter category was named by Vladimir Nabokov, who is still considered either obscene or unreadable, certainly igNobel. If you ever do read *Ada* and understand it, drop me a line. All right. So the publishers are naughty. Right. Now back to business.

Poteri put her cigarette out and said "You feeling pretty good now?"

He smiled.

"Well, the call of the faith healer and all" she said. "We're savin' babies' *lives* here." Her voice was deeper, a parody of Hebner's, with the "here" pronounced as "he-a," which in no way resembled his speech. Miller smiled. "Ah'm jigabooin' back to work naiow" Poteri said, and closed the door behind her. The sounds of the crying babies and thumping restless older children came in.

Freud, that lyrical magician, said "The goal of all life is death." He italicized it, as well he might. Later, in *Civilization and Its Discontents*, when he could have retracted that decision, he simply qualified it. The salvation he offered us was simply our society, cowboys and Indians, a fight to survive. Which as we know in these final minutes of the twentieth century is like offering shark repellent to a man who drowns at sea. Ring Lardner said it differently, and I wish I'd said it first: "I can't live and I know I'll never sleep tonight." King Lear said all of it all of the time. He was always making scenes and I don't read about him any more.

In his journals, Herman Melville—remember him? the guy who wrote that big book about the whale?—said of Lake George, ignoble subject, "foam on beach & pebbles like slaver of mad dog—smarting bitter of the water,—carried the bitter in my mouth all day—bitterness of life—thought of all bitter things—Bitter is it to be poor & bitter, to be reviled, & Oh bitter are these waters of Death, thought I."

Melville's family was disintegrating and nobody would publish him after a while. He was not happy. But look at what Darwin said in *The Descent of Man*—you've heard about our descent?—"But what are we to say about the rudimentary and variable vertebrae of the terminal portion of the tail, forming the *os coccycx?*" Your great ones are always asking tough questions like that. "A notion which has often been, and will no doubt again be ridiculed, namely, that friction has had something to do with the disappearance of the external portion of the tail, is not so ridiculous as it first appears."

Do you know what that means? Of course you do. That the world can wear your ass away, but that you need not disappear. You do not have to vanish.

Other things do. People. But not *you*. Which brings us to the recent celebrated suicides. Berryman:

> I don't feel this will change.
> I don't want any thing
> or person, familiar or strange.
> I don't think I will sing
> any more just now;
> or ever. I must start
> to sit with a blind brow
> above an empty heart.

His Greekest poem: so beautiful! It is a resignation and is celebrated for its heroism. All right. But it is also an invitation to the criticules—V. Nabokov's term—of America to call for sculptors and designate great fame. Because Berryman sees himself as a staring chiseled bust in some library's long hall. Not only does he renounce the power of language, and the ability to use it for overcoming life, death, and related matters: he employs the renunciation of overcoming so as to overcome. He pictures himself as a new Homer, staring blindly in marble forever over the scene of his apotheosis. He uses his dying to wrest from us, who are mired still in time, the admission that he is our best. A brilliant achievement. Renouncing, he really renounces nothing, claims everything, gains through what he says is loss.

All of which, we hasten to add, is brilliant critical reading on my part. Except that Berryman no longer breathes. His family is aware of that.

Sylvia Plath tried blackmail with

> Dying
> Is an art, like everything else.
> I do it exceptionally well.

First of all, her "everything else" is *not* an art: it's what isn't art. Recall Ad Reinhardt's perfect "Art is art-as-art, and everything else is everything else." Plath's poem is interested in the other stuff, the nonart. Dying is not an art, and "everything else" is not an art. It's all a threat that she will die unless we somehow give her what we've failed to provide, by making her dreams come true. Her children are aware of this.

Here is what Virginia Woolf said—not when she threatened to die, but did:

> What I want to say is I owe all the happiness of my life to you. You have been entirely patient with me and incredibly good. I want to say that—everybody knows it. If anybody could have saved me it would have been you. Everything has gone from me but the certainty of your goodness. I can't go on spoiling your life any longer.

Leonard Woolf read that letter on Friday, March 28, 1941. My son was tiny then and I was leaving for a war. Miss Woolf made a reparation with her words: art should make a reparation and *not* a death. It is supposed to go into the obscene shit heap on which our world is built, and come back up with the blackest facts of death, and then build life on this. Who says? I do.

In this time of the so-called artistry of suicide, and revelry about such death, we contemplate these dyings daily. What is it like to be old and sick, or feel old and sick, deserted. To see no recourse—not to write about it, sing about it, advertise it in *The New York Review of Books*—but death? To be, say, an old woman who started life by dreaming, who lived her life by demanding that the dreams come true, who is finishing now by seeing at last that few dreams do—that dreams are meant to be lived in the mind while the body and wit must cope with the deaths of hungry orphans?

What is it like to be the woman who finally feels, but does not understand, such great or tiny truth? And who does not want to live in a world of such wakings, prefers the long winter trance? She has no sleeping pills, has not squirreled them away. In fifteen minutes, nevertheless, she is armed—her husband's razor blades, the kitchen knives, a quart of bleach, packet of mouse killer, the window with its drop. She sits in the kitchen at their wooden table with no lights on in their flat. She is dressed as if for life. Does she fill the tub and cut her wrists? Lie in the oven and cook her goose? Swallow everything and wait to dissolve? What are the aesthetics of her choice? And does she leave a note? To whom? Her husband who failed to be what she wanted? Who never came close, even, to rearranging the planes and angles of the world in order to help her be the woman she dreamed about? The child who fled because families are—write it down, it's a definition—what you finally have to leave?

What should we say to the old woman on Riverside Drive? Perhaps we should say *Wait!* I forget why we should say that. I've lost my own logic. But I think we should. It has to do with love, right? Does anyone here have thoughts on the matter? Maybe it has to do with need. I invite your comments.

You may raise your hands.

Please.

No hands? Wouldn't you raise your hands to keep an old woman from dying? Don't you know what to say? Can't you guess?

Then write down silence in your notebooks.

Write down, in that case, that you side with Freud: the goal of life is death.

Or *choose!*

Hebner came in and looked at him. "Don't you want to get dressed?" he said. "The ambulance should be here any day."

Miller said "I'm fine." Hebner came closer and looked at his eyes, took Miller's wrist in his large hand and, with his fingers on his pulse, looked at his watch.

"You're not too bad" he said. "Can I have the nurse bring something in? You want a cup of tea?"

"Miss Poteri? Wouldn't she want a tip?"

He laughed, and his wide eyes rolled up, he nodded his head. "She'd tell me to truck on down to niggertown. Isn't she gorgeous? She's got a brain like a dinosaur, but if I fired her she'd slap a suit for discrimination onto me so tight, I'd have to close up. Okay. You call me if you get worried, I'm going, I'm behind."

The door banged and Miss Poteri called "You are really behind now, how about it?"

"Jiminy Cricket with garlic" Hebner said. "Take it easy." The usual wave of children's voices rolled in as he opened and shut the door. Miller buttoned his shirt because no one else did it for him.

In *The Interpretation of Dreams,* Freud says "As a general thing, the dreams of a deceased person of whom the dreamer has been fond confront the interpreter with difficult problems, the solution of which is not always satisfying." I offer these as the funniest lines ever written on the subject.

The suggestion that you can't always solve the problems which arise from being haunted might cause you to think that the topic is, oh, *Wuthering Heights* or *A Christmas Carol*. Absolutely not. Nor will we discuss H. L. Mencken, Sinclair Lewis, Willa Cather, Kate Millet, or anyone named Pushkin. Why not speak of our own Christmases, birthday parties, family reunions? Why not speak of our hauntings by *ourselves?*

Once upon a time there was a time when I held the future in my hands. My son, a few days old, was extended, with little left over, from the crook of my elbow to the cup of my palm, in which his head lay. He was hairy, unattractively wrinkled, red—not unlike me today. I remember that picture of us—or recreate it, no one ever caught it with a camera—as I stood in his dark bedroom shortly before dawn and comforted him. Or comforted me, in the darkness, while he calmed from crying out his preludes to hunger. On the wall of his blackened bedroom were some photographs my wife had taken, framed in Woolworth's imitation wood and gleaming in what little light came in the window from the street.

With my free hand I lit a cigarette and it glowed in the glass of the three pictures: for a second, as I breathed that childhood in, there were three different fathers, three different sons. They were extinguished when I breathed out. A tiny fragment of heated ash must have fallen on him, for he cried even louder, then, and my wife came in to feed him, or silence us, to somehow make a disposition, and I never saw us again like that in life though I see us now if my eyes are closed. Is that a haunting? Sometimes I still close my eyes to see us. Is that the dead history of a living man which comes to haunt him, or a man who calls the dead past—calls *death?*

When he was two and wouldn't sleep—he wailed at the door of his bedroom until he gagged—I went inside and sat on his bed and held him in a kind of fat curve, one hand cupped beneath his bottom. That was all a hand would hold, or an arm, compared to what it used to hold. I hugged him and pushed my face onto his head, and he lay his head on my shirt, gripped the cloth with his fist, and simply breathed to the rhythm of my breathing. In and out, the two of us, lulling each other—I, fooling him, I thought, into restfulness; he, convincing me that he was fooled. I sat in a tension of sentimentality, hardly moving except to breathe in the easy sighs of sleep, mourning already that I wouldn't for very long be able to hold or fool him with such tender insincerity. And then, already smarter than I was, he laughed out loud and thumped on my chest.

He grew longer, fatter, was always traveling in the isolation of his own moves and gestures. Children get hard to pin down for a satisfying caress. They get stroked, after a time, on their own terms. Who knows that better than you? I'm not brilliant—and who knows *that* better than any one of you?—but I did learn that. And possibly I feel haunted, like a moist old house infested with rats. All right. But I predict that *he,* the little boy, my son, will one day be haunted too. By the same image I've conjured. He won't remember it, but he'll think some time that *I'm* remembering. He'll regret, in a slimy corner of his own haunted house, not being there for me, not able to be small enough for me to hold. Generous errant sentimentality, though perhaps a work of art.

Miss Poteri said at the door "Ambulance is here. Mister Miller?" He thought about the children's poems from Terezin death camp, what he had said about children as storytellers, Ahab as hero, the Satan of Milton, Odysseus and Penelope as family heroes in the domestic epic, Anna Karenina and Dickens' Carker as people whose sexuality explodes when they make love, at last, to giant fast trains, and Labia Lit, the new feminist pornography, and Borges as second rate, and Rex Stout as the best working novelist in America, and Israel, the Nazis, Cambodia, the Kurds of Iraq, Jack Kerouac, his final lecture swelled out before him, shimmering, like a giant bubble from a baby's plastic pipe, he saw it tremble and swell. Miss Poteri was standing in front of him, saying "It only took an hour on account of they had to do a hemstitch on a P.R. who got laid open by his wife who he was beating up. You get to ride with a pound and a half of sliced P.R., Mister Miller, they figured they'd save gas and take you both at once. Mister Miller?"

She put his tie under his collar and then knelt to fasten one of his shoes. Outside, infants cried and he heard Hebner's deep soft voice talking to the one he examined. A man with a white jacket and a bushy beard was behind Poteri now, and he reached for Miller's arm.

Miller pushed back on the carpeted floor with his feet and held his arms at his sides. "I don't want to go" he said. "Thank you very much, but I don't want to go with you."

The man with the beard said "Jesus Christ, we got a hundred places we could be. Why didn't you *call?*" He said to Poteri "What about the doc?"

She pulled a cigarette out of her uniform pocket and lit it, offered one to the ambulance attendant, who shook his head, then blew out smoke. She said "Screw him. Take the P.R. and I'll take care of Mister Miller. He'll go later, in a cab. Right, Mister Miller? Yeah, go ahead, don't worry about Malcolm X, I'll take care of it." She offered Miller her pack of Lucky Strike and said "You want one?"

"I'm not allowed to smoke. Thank you."

The ambulance attendant shrugged his shoulders, gave her something to sign, and left. When he closed the door the voices of the children went away. Miss Poteri said "Hey. Mister Miller. You want to talk about anything?"

He put his head back against the sofa and closed his eyes. "Maybe later" he said.

But he didn't talk, not after two cigarettes, which Miss Poteri smoked while looking at the wall behind him, and she left to work at taking care of babies. The harsh music of children grew louder outside, and he went into it after a while. There were three doors opening off the short corridor, parents and children were in each small examination room. At the other end of the brightly lighted white corridor was the door to the waiting room; it gave a sense of bulging-in from the heated air and tension and fatigue that pressed against it from the other side. Hebner, in Room 3, was pulling a disposable diaper from a baby who screamed in the back of his throat as if the flesh were tearing there. Hebner looked down the throat and in the baby's ears, threw a tongue depressor across the little room past the baby's mother and missed the wastebasket. The mother watched it fly, and looked at the five or six tongue depressors lying on the floor around the basket. Miller walked in, stooped, picked them up and threw them away, then walked back out to the corridor to listen as Hebner, telling the mother "Hold his arms down, please," took a throat culture by poking a long-stemmed cotton swab down the baby's throat. While the child raged, he pressed the bacteria-heavy mucus onto an agar plate which he put back into his coat pocket.

Miller leaned against the wall in the hallway. Hebner came out, closed the examination room door, and stood in front of him, looking at his face. "Do you teach science at Columbia?" he said.

Miller shook his head. "I just retired, I don't teach anything. I used to talk about books."

Hebner nodded. "How do you feel?"

"Bewildered."

"Not worried enough to go with the ambulance, though."

Miller said "Plenty worried. But I don't want to go to the hospital. Not until I have to."

"And you know you don't have to?"

"No."

Hebner said "I don't follow your logic."

"I'm not being logical."

"You don't want to be followed, in other words."

Miller smiled.

Hebner pointed at Room 1 and said "I'll be in there if you want me. Don't be brave. Don't be stupid. Call me."

Miller raised his hand and lowered it, Hebner reached for the hand but it had fallen to Miller's side. Hebner nodded and turned and went into the examination room. Miss Poteri looked around the corner, from a little office that opened through a window into the waiting room, said "Call us if you get woozy or something, Mister Miller. All right?"

Miller smiled. He leaned against the wall, then walked slowly over to the examination room. Inside, Hebner sat on a high metal stool and looked at a woman in baggy pink slacks and a rumpled off-tan blouse which was coming out of her waistband. Hebner's sleeves were rolled up and Miller looked as the coffee-and-cream and very shiny skin rippled and jumped when Hebner reached his arm out to summon the five- or six-year-old boy—slender, wearing dungarees and sneakers and a blue-and-white striped jersey—who watched the hand, then walked in between Hebner's spread legs and up to his lean belly, where he rested his head while Hebner rubbed the back of his neck, stroked his long bony back. Then Hebner pushed the boy away and carefully looked at the irregular bruise, surrounded by scraped red skin, on his left temple. The wound was about the size of a baby's fist and it was very angry. The mother was saying "So he told him whenever he sees him on the street with another woman, he's supposed to call him uncle instead of Daddy. How can you say that to your own kid? And what happens, he gets in late one night, and Billy 'n me're watching TV, and we get into the usual fight, you know, whatever, the usual kind of thing. And he calls me a whore or something. Right in front of his own kid. So I say something, you know, who're you to talk, running around with two-dollar hookers or something, and right away he figures Billy squealed on him. Which he didn't do. But he don't care, he *figures* it. So he reaches out and belts him, right off of the chair onto the floor. Can I put him in jail? You think I can get the son of a bitch put away finally?" Billy grinned. Hebner gently swabbed his wound and looked at it closer.

Miller walked away, down the corridor, back into Hebner's office. In the dim light there, he sat again on the orange sofa and rubbed his chest. Soon he lay flat on the sofa, one leg still on the floor. He stopped rubbing his chest, folded his hands together over his belt, closed his eyes. When Miss Poteri came in, he opened his eyes, then closed them.

She whispered, as if someone else in the room were sleeping, "You all right?"

He shrugged his shoulders and kept his eyes closed.

"You want me to call *some*one?"

He opened his lips which were very dry, and the sticky sound of their parting sounded loud because the noise of the children was sealed outside the door. He said "I think I'll be all right."

Miss Poteri said "You have an illness, you know. You have to do something if you're sick. You can stay here all you want, but it won't make you better."

Miller said "How's my pulse?"

"Your heart's beating."

"You see?" he said.

She said "Let me call someone for you."

His eyes were still closed, his hands still folded over his belt buckle. He raised his dangling foot up onto the sofa, crossed his ankles, sighed, and closed his mouth. "Later" he said. "All right? Maybe later." Then he said "Do you think Doctor Hebner would have dinner with me tonight?"

Miss Poteri lit a cigarette and, as she blew out smoke, said "He hates white people. You live alone? I'll make you supper."

"No," Miller said, "that's all right. Thank you very much. I'll go home. I'll go home pretty soon. I don't live alone."

Miss Poteri said "Whatever."

Miller said "No, really—"

"Whatever."

The door swung in hard, so that its knob bounced on the wall, and Miller rolled over on the sofa to push himself up with his hands and look over the armrest, like a baby learning to support its head. "Oh my God" he said.

Matthew, in his loose huge jockey shorts, but otherwise naked, howled in the doorway. "I don' like dose bloodshots" he said. His smeared mashed face was stretched for crying and it hid his eyes. "Dey hurt me, Doctor uh—" He held his hands out to Miss Poteri and said "I forgot his name again." He clasped his hands and moaned. The disposable syringe wobbled in his arm and, as his hands moved, it fell away and bounced on the rug.

A little pool of blood formed on his white toneless skin and Miller stared at it. The bank robber's pushed-in face contorted again, and Matthew waved his hands at Miss Poteri and moaned.

Miss Poteri said "Son of a stupid bitch" and put her cigarette into the tray on Hebner's desk, walked a few steps to Matthew. Miller looked over the sofa's armrest and didn't move. Miss Poteri said "Son of a bitch." Doctor Hebner was in the doorway behind Matthew, then, and he put a hand on the back of Matthew's neck. The fat short neck ducked forward, the almond-shaped head moved back and forth, and Hebner's hand dropped down through the air and hit his own leg.

Matthew said "What's his name? I forgot his name again! I don't *want* a bloodshot! What's his name?"

Miss Poteri walked closer, and stood before him, and then said "Goddam stupid son of a bitch" as she put her arms out and hugged them over Matthew's shoulders and pulled him against her body.

Hebner hit his empty fist against his leg. Miller leaned his face to the orange armrest and closed his eyes. Miss Poteri said "Son of a bitch. Son of a stupid bitch." Matthew brayed.

Miller said "I'm all right. I feel all right."

THE
SENTENCES
1975

Claire has watched the postman force our mail into the brass box, and Claire has torn it out. Saying "Good morning" so that, if it depends on her wishes, he doesn't have one, she has come upstairs from the little lobby on Riverside Drive to drop the day's delivery before me in the kitchen. It isn't a happy time. It's cold, it's almost Christmas, the oil-producing countries of the world are offering a sly exchange: Israel for energy enough to burn our lights and move our cars. Things are flickering.

But they still deliver the mail, and I evaluate what comes: ad cards, newsprint catalogues of bedsheet sales, Marboro remainder notices, *here;* bills, which I always pay on time, go *here; here* for the letters from Columbia, which retired me early, may it sink—it probably will—into the angry brown skins which surround it; and *here* for the personal mail. Personal, in turn, gets subdivided into those I love and those I don't. And what do I see but a letter from Harry? I grunt in my chair, my serenity shatters; I tear open the envelope to see what's the matter.

My fat wandering thirty-six-year-old sends me his usual war between the grace of his handwriting and the frenzy of his exclamations. "Merry Christmas" I say to his letter. "In two weeks it's Christmas, and what do you have to say?"

Two-strike. Buffalo Woman. Hard Robe. Man Not Afraid of Pawnees. He mourns their passage from the world. He says *Real Americans* about them and then, a couple of sentences later—he is careful about his punctuation throughout—he tells how Chinese some of them look, or how Russian. He says *Americans* again. *But they're only in pictures now. Everyone is only in pictures.*

It seems he's visiting a friend who teaches film—I envision a bearded man lecturing in a large room, at every desk of which sits no student, but a metal reel of 35mm. movie film—and, shivering in sentimentality and the absence of heating oil in upper New York State, he wanders the college library *waiting for something in the stacks to fall in love with me.* He looks at wide thick folders of rotogravures—sepia, shiny, set in shadows on the long skies and bunched-muscle mountains of America: oversized pictures of the Indians who had posed so patiently for the white man and his box on sticks. As if they knew that they would disappear and he, who had come to them from a race which specialized in disappearances, could somehow therefore offer them (and guarantee) their single hope for permanence in the coming world of trains and endless wire.

Cornplanter. Kaibah. Weasel Tails. White Eagle. He sits in the high library, the winds hum around the corner where he is and wrap him in cold, the humming high fluorescent lights are like a shining wind, and he looks into their faces—*Wah Ti An Kah. Hard Robe. All the Waters*—and they look back. They speak to him of being beyond one's touch yet touching. He holds his fingers on the margin of the cushiony paper, and he looks in.

The second or third day there—*Even in libraries. Every-*

place. Something comes and gets you—he sees that the tissue paper which has separated each picture from the one above it in the portfolio pile is eating the portrait it's protected. *Acid in the paper,* he writes. *It isn't anyone's fault, they just made the paper like that in 1905, 1910. That's the way they lived.* Each fold and crease of the tissue paper has been printing itself a little, year by year, on the face of the Indians whom white librarians thought to protect. And late in the century, a little before Christmas, those seamed dark faces bear the scars of protection, their coming erasure. *Nothing lets anything alone,* he writes. I say "Amen." He adds some facts about his broken car.

I see him in the stacks—tall and fat, his hair long and ragged. He is so ugly and alone and such an easy target for the weapons of the world to hit, I close my eyes to refuse their tears, my nostrils sting, I curse us both.

I see the stacks. I have to picture Burgess-Carpenter Library at Columbia, may their sit-ins broil like old wounds, may their Early Retirement for Undistinguished Professors leave them only infants in long flopping trousers who giggle and tell their students "Dah! Da-Dah! *Boof!*" Old paper in turn-of-the-century editions quietly stews in its high acid content. And then—who else's?—the acids of Jonathan Swift in a nearly useless edition begin to bubble and steam. I put Harry's letter away, past the stack of advertising mail, so that I cannot see what I've nearly read—his long neat crazy comments on his memories of Christmas in Brooklyn, when he was little and I was at war in 1941, when the world was a thought one could still entertain.

I wear my raincoat because the retired Nazi who owns the building has clearly co-operated with our President in conserving fuel. Today he will permit us to serve the Fatherland—always *fathers!*—by living without heat or hot water. His gawky janitor hides from the tenants and apologizes when he's caught. He's only obeying orders. Tomorrow he'll distribute stone soap.

And the acids of Jonathan Swift are now so strong, and so hungry, they are digestive juices, scoring the gray metal bookshelves in the stacks, splashing upward onto Laurence Sterne's

Life, onto all four parts of *Tristram Shandy,* splashing north-by-northeast onto *Travels with a Donkey* and *The Master of Ballantrae,* north-by-northwest onto Matthew Prior's *Literary Works.* Ruskin is pitted, and Arthur Symons (due west) gets burned. But down below, as the acids foam like tides, *A Tennyson Dictionary* ("Roä: Name of a dog") is eaten whole. *Thackeray: The Critical Heritage* and Dylan Thomas undergo an enzyme digestion, as does *Great Reading from Life* and *The London Omnibus.*

Oscar Wilde survives. So does *The Age of Extravagance.* The gray-white marble-patterned linoleum floor is spattered with the burned-in acid-printed flesh of books. Shelley, on the tall blue shoulders of Edmund Spenser, 820.81/Sh over 820.81/Sp, squints down and wipes his nose, paws at his hair and writes: "continue to believe that when I am insensible to your excellence, I shall cease to exist."

I add, for Harry and myself, "Well—bullshit." Just as he has made the vanished American Indians into objects of his own sensibility, I have made him into one of mine. We have done it endlessly, Harry and I, creating ourselves for each other, and creating each other for ourselves. We do it to the world—we're sentimental, and we hide from history by making it our own—but the world will not reply in kind: it gives back data, *facts.* How long can you hide from a fact? A coldness? Angina and obesity, my arctic and antarctic poles?

Harry, how long can I hide from what you have written to me about Christmas thirty years ago? It's like never again reading *Jude the Obscure* to never see again those children's bodies hanging by box-cord on the garment hooks of a shabby flat in a university town. See: I see them even as I burrow into my coat. Like never wanting again to consider whether because I was not *bar-mitzvah*'d I'm a Jew. Like wanting never again to read the journals of survivors of the Holocaust. Never to know of gang rapes by twenty-year-olds in dark moving cars while the high school cheerleader shrieks at the vomit which gags her. Never to see the serpent's tail of the rat which gnaws at the cheekbone of a child in a room uptown. But history seeks us out.

My roving son returns in my coffin-shaped mailbox to tell me of facts. He reads in a library idly and facts devour him. My heart will squeeze in my chest and hold back my breath to tell me facts. And facts—the stink of *pogrom* and international oiliness—come into our rooms at Christmastime to tell me of the vampired oil of the earth, of cries for blood, and how the world is growing cold and, like tired light in greater darkness, trying to go out.

I reach for my nitroglycerine. I reach for Harry's letter. Claire comes back in and I swallow the capsule, put the letter down and wait. I see the acids from the Print Library wash through the plumbing pipes and into the floors below, to eat at History downstairs. I hear a librarian screech "The Indians are swallowing the Jews!" Harry, I am using you again. When you try to write, do you use this damaged marriage, fact you flee, as I use you?

Claire stayed young-looking long. Her hair stayed dark, her skin stayed smooth. I aged when I was forty, and have listened to my heart for thirty years, urging it to treat me kindly, praying to the stupid fatty repetitive god. Now Claire sees her throat gone slack, her chunky shape gone rounder, and the skin of her upper arms gone soft as dough and pendulous. I hear her lips purse when I read *The New York Times* obituary page. Often, when I say out loud the name of someone who's died—a public fact—she pretends not to hear. She interrupts with a comment about a translation she's doing, or a book we haven't read. As I go through the Sunday *Times* she idles like a badly tuned car, anxious, wanting (she never says it aloud any more) to be off to the Museum of Natural History, or one of those endless cheerful Audubon Bird Club walks, or a serious movie at the New Yorker (all I need is sadness to *see,* in believable motion and color). Instead, we sit. I think and adjust my body to the signals from my heart. She hates our life.

So Claire comes in and makes a late-morning pot of coffee. I say "A letter from Harry. He's visiting upstate with what's-his-name, the Ph.D. from N.Y.U.? It's called 'crashing' if you bunk in with someone for a while."

"I'll bet it's called crashing. Crashing bore."
"Who? Harry?"

She addresses the drip coffeemaker: "I wish he'd grow up before we die."

"As one who has presumably grown up before I'm dead—I know there are arguments to the contrary—I think I should say for fairness' sake that it's not as much of an advantage as it's supposed to be."

"Thank you very much. Is Harry all right?"
"I guess. As all right as he usually is. He's sad."
"That's not news."
"I guess everybody's sad."
She says "I guess everybody is. There's no return address?"
"No."

She goes out the door and back to her desk in the dining-room-and-library to wait for the nearly boiling water to filter down. I think of Jonathan Swift's voracious acids filtering down. She will translate a sentence of Russian political prose before she returns. I reach for Harry's letter as the sun we can't see from our window wobbles coldly up toward noon.

He writes *I've been keeping diaries. Journals of dreams. It helps me stay straight. You would call it narcissistic. It ends up, I write down as much about my wakings. A great delight to write. Does the world need another set of diaries? Someone could do a set of Collected Letters from Underground. Underwear, anyway. Jesus, it's cold here, Dad. See, they get 15% less fuel oil for heating than they got last year. Last year it was a mild winter. Ipso dipso, they're in trouble. How is it in the hot city? When they stop shooting that blast of hot air up the front of Orbach's near the doorway, I'll know they know in New York about energy. Or the SPRY sign in Jersey we see from your house. Our house. Sorry. Anyway, I've been writing things down. Do I need a new paragraph here? I ended up writing about when I came home from college in my sophomore year saying you and I shouldn't kiss each other any more. Just shake hands. You were so nice about it. Because you liked hugging me. I always knew that. But you shook my hand as if I* were seventy-five years old. I think you were beautiful. When I came home later on and kissed you by mistake, you didn't say anything. We kept on kissing after that. Remember? I wrote it down. I dreamed about you walking up the attic steps in Brooklyn. Walking up very slowly and not coming down. It felt like the year we sold the house, my senior year in school. When Mom did the Turgenev for Harper and Row. I called up from school to say I couldn't come home as planned. You said good. I said why. You said Mother is under considerable strain with this assignment. I remember that. Considerable strain. Mom, if you read this, don't think I'm being uppity. Just, I date the major part of your marriage's destruction from then. Destruction's a dumb word. No one did anything on purpose. Did they? Things got bad. Considerable strain. You worked later in school, Dad. In those days no one stayed late at Brooklyn College. It was like a vocational school. Midwood High School had a better reputation. I don't know why I'm saying this. Apologies to the marriage.

I say to the coffeepot "No apologies necessary. We stopped fucking before we started to stop talking. It should have been the other way around. Get it up, *then* shut it up." Like my son, I'm a vulgarian.

Claire calls "What did you say?"
"Did I disturb you? I'm sorry."
"What?"
"I said I'm sorry."
"Oh." Then: "Why?"
"Nothing, honey."
"What?"
"Nothing."

I button my raincoat because the chill in the apartment is a coldness now, and my hands are aching, especially the one I hurt in the war. I think of that, of Italian mountains, of field hospitals—how cold I always was beneath the bandages—because I've seen how toward the end of his letter Harry has written what I fear so much to read: about his tiny childhood which I missed while he was beautiful and barely handled by the world and I was in a bed on the shore of Lake Como, and every other morning the nurse with a face like an old dog's would come in and break the calcium which deposited in my fingers; I cried like the child my Harry was. Perhaps that's when we were closest, and never knew. I wore my wedding ring on a thin gold chain around my neck. I clutched it when they broke my fingers. And now, in my marriage again, I clutch my wounded hand when Claire comes in, wearing a quilted down-lined jacket and looking like a khaki pigeon.

Pouring coffee she says "It's getting colder."

"Perhaps it's that thin chilly prose—"

"Very funny. Except that man is a *hero*."

"Yes."

"He's—dynamic. He and his wife live in *fear* in Moscow."

"I know."

"Except they're not afraid."

"How can you live in fear and not be afraid?"

"I mean they live in a climate of fear. They live in *spite* of their fear. The intellectuals there have made them a rallying point. Well, you know: you taught his first novel."

"Yes. An awful book with ten or twelve fine pages. I'm not sure he can write, Claire."

"He can write. I read him in the original."

"I'm reading Harry in the original."

"Meaning what I'm doing now is a waste of time? I should be worrying about my prodigal son? Well, listen: I worry. I worry plenty. I worry *sick*. When will he worry about me? About *us*? You know all this."

"Honey. Honey."

"Don't call me Honey."

"Claire."

"Thank you."

"Well—"

"When will he worry enough about us to come home? To write a letter to *me*? One man smuggles manuscripts out of the worst police state in a *world* made up of police states. Another can't drive his car a couple of hundred miles—"

"Honey—Claire. I think Harry is locked away the same—in his own—"

"No! It isn't the same, I won't *hear* it's the same. A family isn't a government, that's so facile it's disgusting. Disgusting."

"He loves us."

"Let him show it."

"Everyone can't always show these things. Are you crying?"

"I'm pouring coffee in the sink because my stomach hurts."

"Are you sick?"

"I'm sick."

"What, Claire?"

"Nothing."

"Is it the cold?"

"Yes. It might as well be that."

"Claire—"

"Excuse me."

Excuse us all. God bless us one and all. We didn't say it once again. Whatever it is, all the things we know it is, we didn't say. Which is one of the reasons, or symptoms, perhaps, for your writing us from far away. Listen to my heart bang, Harry. Tell me what I dread: these facts.

This happened a while ago. On Route 17, the car banged and shook and stopped. I had to get towed. It turns out some crook sold me watered gas. I guess we have to share our great American energy crisis, huh? Pass it around. So they're taking my carburetor apart or something and I'm hanging around the gas station, waiting. The radio there is going and you know what I hear? This old record. Bing Crosby singing "I'll Be Home For Christmas." Jesus, Dad. I walked around. I looked at the motor-oil cans. I started to cry. See, I remembered the living room in Brooklyn. That weird rug with the fruits and vegetables on it. This huge Victrola near the fish tank and the upright piano. Mom used to play this album of 78's, Bing Crosby singing Christmas carols. This deep voice saying I'll be home for Christmas, just you wait and see. And I was always sure it was you. I would say "That's my daddy. That's my daddy." I don't remember how I felt when you didn't come home as promised. I don't remember how I felt when Mom explained about your being in the war. We had a picture of you from when you came home on furlough. Before you shipped out for Europe. You were so big! You were such a handsome soldier, Dad. Me and my energy crisis. Watered-down gas. You singing to me through the radio. Me knowing I wouldn't be home for Christmas. Who skipped out of my turn in a war. The feel of your whisker bristles when you hugged me. Hugged Mom. And she'd say ouch and I'd laugh. When you came home I said to Mom "Let's pretend our husband is dead." She beat the hell out of me. She was right. I mean it was fair. Little Oedipus of Brooklyn. You gave me a rabbit doll. We called it Daddy Bunny. I used to beat hell out of it. Mental health. I threw it out the attic window. Three floors down to the dog run next door where they had that incredibly savage chow with his purple tongue. He tore the doll to pieces and I laughed. Mental health. I miss you. Goddam Christmas sentimentality. I don't think I'll be home, Dad. Things keep happening.

I call "Claire! Claire!"

She comes in with her expression of fear and distaste. She worries for my health. She hates the idea of my leaving for good (we never say "die"). I have kept myself alive, I haven't left her alone, though I sleep in Harry's room and my balls are flaccid and fat—shriveled now in this coldness like an infant's in his sac of softened walnut skin. We speak of love, not making love, and we eschew the vulgarest pictures and books—or decide they're not about sex, in fact, but love. She says "You're all right?"

"It's too cold in here. It's too cold!"

She goes to the hutch for a sweater for me, and my eyes wetten at how quickly she is Claire who depends on my presence, and not the speaker of secret tongues.

I say "Things keep happening to us! Dammit. They owe us *heat* up here."

I refuse the cardigan she offers and go for the living room, its door to the hallway, the three flights of stairs and Marciak, the young superintendent, with his tape recorder cassettes, his job he doesn't do.

Claire says "What did Harry say? What did he *say?*"

"He loves us very much."

"Where's the letter?"

"No—it's mine, for me. Later I'll tell you about it."

"What's going to happen? What's happening?"

*"Some*thing is happening."

I close the door behind me, leaving Claire inside. I'm in the battle zone—short corridor, marble floor, dim lights, and a locked metal door—and it is colder than our rooms. I say "Hubba-Hubba" and I laugh. I sit on the first marble step and hold the metal banister rung. I lean my forehead on it in the dimness, then pull my head back because it burns. I breathe hard, then harder. The pain across my chest is no surprise. I pant, but I speak between the gaspings: "More. *Heat."* No one answers in the house, its cold and darkness are undisturbed. The ice stone stairway disappears into black below. I say "Heat" but can't hear my words. Then our doorlock ratchets open and my heart jumps. Claire runs to the stairway, a waddling small woman on loud little heels.

She says "What? What?"

I lower myself in the darkness. My arm which clutches the banister bears me slowly down. In Italy I climbed with crampons and a rope belayed by a six-foot-five-inch Indian who had worked at Abercrombie & Fitch and who killed a German sentry with an arrow through the throat. Now I sally forth to ask for steam, belayed by a small intellectual pigeon.

Claire pushes the gelatinous ampule into my hand and I know what it is and lay it on my tongue. Her terror hangs around us like a phosphorescence, two creatures from a mountain cave limping wounded into the world, glowing with our strangeness or despair. *That's my daddy. That's my daddy*.

In Italy, on the mountains, reading what tattered mail had come through, I would offer Harry's photographs—every letter had a new one—and read aloud his language which Claire had transcribed. *This is my son.* The ground was so hard and the mortar fire so vicious, we'd scrape at the earth and lie down like frightened children when they shelled. The fat men put their helmets on their ass, the slender ones guarded their brains, we had 80% casualties anyway, and in the field hospital before they shipped me back to Como I dreamed of being roped down the face of a white cliff in the brilliance of a spotlight while below me serpents in helmets heavily rolled their coils. I had no carbine, no belt on my pants. I cried and cried and an orderly from the Negro division would wake me up and say "You are bound for *home*, Sergeant. Don't you scare us all with your crying, now." I came home. I cried anyway and anyway came home. Death owned everything, I thought. Claire hisses behind me of life. And now the world is a slum, a lake of excrement. Still she hisses life. *That's my daddy.*

"Rest!" Claire says at the first floor landing. "Rest!"

"In. Peace."

She whispers "Catch your breath. At least breathe. If you've time for jokes, you've time to *breathe*, haven't you?"

I stand. I breathe. My chest pains don't decrease, but the squeezing does, and I know that on the first-floor landing, in the marble of the corridor, blue-gray light of the entranceway window leeching heat instead of offering it, I will not fall and die. We shuffle to the blacker end of the hall, beneath the stairs, where coldness lies on my skin. I say "Knock."

"What?"

"Door."

"God. You can't even knock on the door, and we're here to demand? How will you get back up? How? You can't even *say* it! How will I get you upstairs?"

But she knocks, there is nothing she can do but obey. We are married by extremity, she can only be alone unless she closes her eyes in darkness and beats on the blind door. She must be praying that the arc of this action will carry us back to where we've been for so long. I know that she is weeping as she slams her little fist against the door.

And heat pours out like long-dammed water, golden light

rolls over us, as Marciak in fawn-colored corduroy pants and a red flannel shirt calls over the toodle-and-peep of Vivaldi that we should come in. He shuts the door behind us, and I breathe in heated air. Claire unzips her jacket so as not to catch cold. Marciak, crouching at his tape recorder to lower the volume, says "Professor! I'm honored."

Fat bright cushions splatter his floor, a red-and-white pan crusted with orange sauce weighs papers down on a folding table. Near a yellow chair an electric heater glows, and another shines from a corner where a white bookcase is packed with tall serious clothbound books. To our right, on a yellow-covered daybed, two books lie open, face down, like sick people in the glow of Marciak's third brilliant heater. He says "Professor?"

Claire still stands at the door. I come forward, look into his thin white face and dirty glasses, his long dark hair. I say "Breath. Sorry. A minute."

He says "Is he all right?"

Claire says "No."

"Well, should I—"

"No," Claire says, "just let him catch his breath."

I say "There."

Claire says "No. At least *wait*."

Marciak says "Wait. By all means wait. I'm not going anyplace, don't rush. Here—sit here, near the heater. Please: sit."

I stand, and so does Claire, so he does too. I no longer feel my chest, I am elevated over it, feel that I could move my head and see my body's out of sight. I look across the room instead, at his stove and little refrigerator, at a Picasso cut from the cover of the Sunday *Times* and taped above the sink. When I speak I'm no longer breathless, but my ears feel clogged, my voice floats up to me like smoke. "Are we disturbing you, Mister Marciak? I'm afraid we feel unsettled."

"What, the heat? Hey, I don't blame you."

"We're cold."

"Hey, I know. It's awful, isn't it? I'm really sorry. I had complaints all morning. I can't *do* anything."

"Who can do something, Mister Marciak?"

He looks at the floor. Claire holds her hands before her, pressed in against her stomach. He looks like a child, he *is* one. He makes me miss my students.

Offering information as a little boy might offer angry parents his newest drawing, he says "I know your son." He smiles at me.

"My son?"

"Well, I don't really *know* him. But I met him. A girl I know, she matriculates? I'm just in General Studies. This girl took me to your last lecture, the one everybody packed into the hall for. It was really nice, the way they all came in there. See, *she* knows your son. She introduced us. We shook hands. I took some really amazing notes off of your lecture. Amazing."

"My son?"

"Yeah, Harvey. No. *Harry*. Harry, right?"

Claire says "Our child is named Harry. Yes."

"That was the end of May" I say. "I gave the lecture last May. And he was there?"

"That's what he came for. He came down to hear you."

"He came here in May for my lecture."

"Right, May. Listen—can't you sit down and let me get you some coffee? Beer? I've got beer. How about some hot tea?"

"I didn't know he was here in May."

"Oh."

Claire says "He hasn't been home."

"No," I say, "that's wrong. He was here last May."

Marciak says, after we've all been silent, "I could have bet it was a tenant when you knocked on the door. You're the fifth this morning."

Claire says "I'm not surprised. It's disgusting. Especially with the rent we pay."

I have forgotten to fight because of the gift I've been given. I say "Yes, couldn't you manage to get us some heat?" I don't care.

"Professor, I can give you one of these heaters. That's about it. They never delivered the oil. I called up the landlord and he called the trucker, but I don't know what's what. The fact of it is, I guess, they just don't have enough. I don't know. I really can't help. Would you like one of these heaters? I mean, it would be a real pleasure."

"They're wiping us out" Claire says.

My voice tells her "It's only a possibility. Now I'm not sure."

She says to Marciak, devoted minor character, or me, "In Asia, any part, all the writers worth reading are locked in jail. In South Africa they torture the priests. The French government bugs editors' offices. In America, Nixon's a Mafia crook. Russia's out of the question. We're selling Israel to Egypt for oil, you can't be a Jew anywhere. And meanwhile the old widows in England die of exposure because they live alone, they're too weak to crawl from their beds across the floor and make a fire, so they die."

Marciak says "Yes ma'am. I know. It's all very—lamentable."

"Lamentable" she says. "Look at Pakistan, for God's sake. Or New *York:* we pay four hundred dollars a month for this—this middle-luxury apartment, it isn't a palace, but still: what do we get?"

Marciak says "Cold?"

"Exactly. Cold. So what can we do?"

"Ma'am, won't you let me give you one of these heaters?"

She says "We don't take heat from students, from the young."

I say "It's such a coincidence. That you work here. That you saw Harry. That he was here in May. In the class. To see me."

I hear the smile in Marciak's voice, his relief: "You see, ma'am? *Some* things work out."

She says "When I see him, I'll know they work out."
I say "Well he was here."
"Merry Christmas" she says. "Wonderful."
I say "Good-by, Mister Marciak. Thank you. Good-by."
"Claire says "What are we so *thankful* for?"

"Excuse me, ma'am, but I don't think the professor should walk those steps now."

I show him my teeth and nod my head, though I don't feel it move. I say "Thank you. Your—thank you."

Claire says "That's his bluff and hardy voice. It means we're going to be heroic. Whether we can breathe or not. One of our mottoes is that a little cardiac insufficiency never hurt anyone."

I continue to breathe. Marciak speaks, and Claire does, and so do I. Light moves and then the darkness falls upon us as his door blinks shut. I shiver in the marble hallway. "All right" I say. We move along the floor, Claire holding my arm, and we turn like slow wheeling fish in an icy sea.

Claire says "How *will* we get back up?"

"What?"

"How will we get back?"

"Oh."

"Yes" she says.

"I never promised we would."

"Please" she says. "Don't *die*."

I say "No. We'll discuss it on the way."

NEXT WEEK I LOVE YOU 1976

Shortly after his father dies, the man named Harry who will soon be middle-aged drives to New York and has a discussion with his mother. When he talks he avoids the past and evades the future. She treats the future as intolerable; the past is heavy in her lungs and eyes, a caustic smoke. We stare, we are superior—we speak of them but they can't reply or even imagine us—so we begin *above:* drop like a spider on its drag line through roof and two floors of rooms and through a patterned white plaster ceiling, fast as the breath one draws before shouting, into the kitchen of a flat on Riverside Drive.

It is June and muggy, and Harry sweats in his short-sleeved shirt. It hangs on him where it doesn't stick with perspiration, and his face flesh hangs, around the eyes and mouth especially. He's a fat man grown thin very quickly, and his body doesn't become him. His mother is small and chubby, her sleeveless jersey is purple and old-fashioned in design, her poplin skirt is longer than the skirts worn by women who read the magazines. They are drinking hot tea and their faces recoil from the heat which they bend to.

His mother purses her lips—the lines of that gathering-in look as old as she does—and Harry rubs his mouth. She says "Well?"

He says "I just thought I should come."

"Obviously." She drinks her tea, she gives nothing.

He says, as if he has to, "Is there anything you need?"

Her eyes flood, she shakes her head. Although her voice quavers, she forces the words through currents which jostle them. We move in closer. "That is so far beyond answering to." She stops. "Did I say answering to?"

He nods his head.

She nods back, looks at the dark square coffeetable protected from their cups by a heavy sheet of glass. She is on the sofa in its yellow summer covering, he on the pinewood rocker opposite. She says "You can understand how upsetting this visit is for me."

"I'll leave. Do you want me to leave?"

"Don't sound so injured! You, of all people, have no right to sound injured! My God."

"I'm not injured" he says. "Never mind. But don't use this as a chance to get even with me for my life, all right? I didn't come here to surrender it. I just wanted to see if there was anything I could do."

Her lips are still pursed. She puts her cup and saucer on the glass sheet—they are reflected in it: her hand reaches up to her cup from the table as her hand descends to the table with her cup—and she holds her fists side by side in her lap, pressed against her stomach. "Well," she says, as if by agreement now their tone will ignore what their language says, "I really don't know what you can do." Her words are gliding, smooth. "I telephoned the doctor when I couldn't wake him up. I listened to what the doctor said and I called an undertaker he suggested. I talked to the men when they came for his body and I went to the—" he is putting his cup and saucer down on the glass, and his hand meets his hand in its mirror—"crematorium and I watched him get burned in a box. That's all done, you see."

We are upon him—the beaded upper lip and forehead, the moistness of his eyes, the unsteady hands, the wet wrinkles of his shirt and trousers, his right foot which taps and taps in a single same rhythm. He lights a cigarette, holding the right hand with the left because his hands are almost powerless. The match goes into one of four ash trays on the glass sheet.

His mother says "Please don't smoke."

The cigarette, folded and mashed in the ash tray, won't go out. He has to rub it on the ash tray many times, then wipe his fingers on his pants. He coughs, says "There was a reason I didn't come to the funeral."

"There was no funeral" she says.

"Well I couldn't have know that, you didn't tell me."

"I couldn't have told you. I didn't know where you live. I still don't know. You never told me."

"I read it in the *Times*. I read a little paragraph in the *Times*. You didn't even put in the book he wrote."

"It wasn't a book, it was a little thirty-two page catalogue of Columbia's holdings of an utterly worthless writer's manuscripts. He wasn't proud of it."

"You decided."

"I was the one making the decisions. But his sister, if I have to defend myself, was the one who wrote the notice for the *Times*. A very close family. Whenever one of them dies they feel close."

"I didn't ask you to defend yourself. I'm not attacking you."

"Thank you."

"I'm defending myself. Why I didn't come to the—burning, cremation. I didn't know he was dead. I was worried about him, I was thinking it might be soon, but I didn't *know*."

As if she is empowered to cross-examine, she asks, abruptly, "Would you have come? If I had had the consideration to tell you about your father's death?"

And as if he is compelled to tell the truth, he says "No.

I wouldn't have had the right to. I left when he was alive. Why come back because he's dead?"

"The *right?* Is it a privilege to look at a corpse?"

He says "Sometimes a hypocrisy."

After a second, and then another second, she says "Perhaps you're right. You should know about hypocrisy."

"Thank you. This visit, you mean. You mean this is hypocrisy."

She closes her eyes and she silently cries, the agreement about their tone is broken, it is just another treaty. "I don't know what it is" she whispers. And then, almost blubbering, her head moving forward, wanting to fall to the glass before them and between, she whispers "It would have been nice for someone to be here—"

He stands and she sits back, expecting something sudden, maybe violent. He softly moves, with a fat man's excessive wary grace, to other parts of the room. At another table, he touches an electrified toll lamp. He picks up a book and puts it down. In a far corner, he opens a hinged wooden box and closes it, moves on. She sits completely still, her back is to him and the rest of the room, she faces the door of the apartment which goes to the outside hall. We draw back to an opposite corner and watch them both—her stillness, his wandering—and the silence is a language which they listen to. At the floor-to-ceiling bookshelves along one wall, he stands with each hand on a different book; he takes one out, looks at it, replaces it, pulls out another, puts it back. He moves on—to a blanket folded on a cane-bottomed chair, to flowerpots on a white sill, to a newspaper open on a wooden chest against the wall beneath the window. He says "What did you do with the ashes?"

We are close upon his back now, as he looks out the window to the yard behind the building which is littered and septic. Her voice comes from behind us, it is not distinct, a pale sound. "I brought the urn home" she says. "Then I telephoned a messenger service and a young man with a lavender necktie and sneakers took it to your aunt's house in Brooklyn."

"Is that expensive?"

"What *difference* does it make?"

He is moving again, and at the door to the kitchen he picks up the telephone receiver, listens, and replaces it. Then he picks it up again, puts it down, moves on, to the other wall with its white-painted fireplace and the white-painted plywood board which blocks it up. He moves the little vases on the mantel. He is signaling confusion, saying that he requires guidance or a warning, hints. He doesn't look at her. We see her turn to watch him, then turn back. She pulls her skirt hard over her knees to signal that she no longer has to provide anything.

"Why is your telephone disconnected?" he says, still facing the fireplace.

"What?"

"Your phone doesn't work. I just picked it up and there isn't a dial tone. Are you short of money? What happened to his insurance? Didn't he leave a will?"

He doesn't see her shake her head and let it fall into her hands. She doesn't see him lean his forehead on the mantelpiece and close his eyes. Through her fingers she says "You want to give me money."

"Look. I'd give *anyone* money who needed it."

"Thank you" she says.

"You're picking on what I say, you're trying to fight. You don't really need that now, do you? If you're short of money, why don't you let me—"

"*Shut up!* You're so stupid, do you know that? For a bright person, you're stupid. Even after so long, I thought maybe I knew a little about who you were. It's a common mistake, it must be. Mothers must make it all the time. They're stupid too. Because just because I wiped your ass when you were little, why should I *know* you? And why should I *think* that you know me? Do you believe I'd take a cent from you? If I were starving? If I were *dying?*"

He puts his hands in his pockets and walks around the room the way he's just come. She wipes her eyes and nose with her hands and rubs them on her skirt, rubs her skirt. He stands near the telephone and, with his finger in the third hole, works the dial back and forth.

"I'm sorry" she says.

"I know you're angry. It's all impossible, I know that. I knew it for years. Years. That's the way it is."

"You probably meant well. You probably didn't mean to insult me."

"Any more than I already have, you mean."

"Yes" she says. He nods his head, works the dial. "I had the telephone disconnected because I didn't want to talk to anyone who called about him. I didn't want to listen to them. I didn't want to have to say anything to anyone." She stands up, sidles from between the sofa and table, walks across the room and out. We follow her from where we are, then turn to watch him watching her. He walks back to the fireplace and past it, to the apartment door. He waits here. She enters again and, with her fists side by side against her stomach, small in the doorway, she says "I was afraid that you would call." The weeping starts again, she says "I'm sorry" and turns around, leaves.

The doorbell buzzes, he looks at the doorway across the room which she's left. He pushes his glasses back on his nose. Someone knocks at the door. He calls "Should I get it?" He waits and then he opens the door, steps back.

We look over his shoulder and down at the little woman with beaked nose and huge pendulous throat who wears a sleeveless patterned housecoat. Her upper arms are enormously fat and loose, they wobble as she walks in, saying "I thought so."

Harry says "My mother's resting, she isn't feeling well. Can I help you?"

"Can you help me." Her voice is harsh and low, she carries an unfiltered cigarette which she draws on as she uses the other hand to close the door. She walks toward him, forcing him backward, toward the rocker. She draws again on the cigarette, then says "Feeling well? With this"—she motions with the cigarette at him, the large empty room—"she's supposed to be *well?*"

"I'm Harry Miller, I'm her—"

"Don't say it. Don't call yourself a son, you'll be doing me a favor. I've got sons, I know about sons. Don't say it. I've seen pictures of you, I would know you to see you. You lost weight."

Harry steps backward again, moves around the rocker so that it's between them. He puts his hands on its upper rung and nods, says "Hello."

"I am Mrs. Dickstein who you never heard of because you don't know your mother's life. I heard you in the hall and I watched you through the peephole. I said 'This will be unhappy, for her it's too late for a visit. Can't be happy.' So I waited to be decent and polite and because you never know, it could turn out different. But I doubt it. Am I right?"

He smiles as if she is an ally, she has named a nameless situation. She doesn't smile back. Like a small child entertaining guests before his parents come in, he says "Hello, Mrs. Dickstein, how do you do?"

She walks to the glass-topped table and puts her cigarette out in the ash tray. Her arms shake as she reaches to her housecoat pocket for a pack of cigarettes and matches. "How do I do? I'm a widow, like your mother. We are a couple of old widows. When we go shopping together, people look at us and they wish for us that things would be cheaper. We walk together slow, with shopping bags."

From the far doorway—we turn to see her—Harry's mother says "Belle, this is my son."

"Don't say it" Mrs. Dickstein says.

Harry's mother comes closer, and they stand, waiting. His mother says to him "This is Mrs. Dickstein."

"Yes" he says. "She's been telling me about my derelictions."

"What derelictions" Mrs. Dickstein says. "I *know,* that's all. So I told him." She faces Harry: "Am I right? I know who you are, and I know what's a son, and I know what you're not. You are not this lady's son. That's all." She puts her cigarette out in a different ash tray on the coffeetable.

"We were talking, Belle" Harry's mother says.

Mrs. Dickstein says "I wouldn't have guessed it. If anybody asked me, I would have told them 'She's in the bathroom, crying, and the young man who is so full of duty and love he couldn't come when his father died is in the living room killing time.' But maybe I'm wrong. You know me well enough by now, all you have to say is good-by and I never was here."

"I'll call you later, Belle, is that all right?"

"Your phone wouldn't work unless you used it to hit someone, remember?"

"I'll come over."

"You want me, you call me. Stick your head in the hallway, even if you worry too much about your dignity. Shout. Scream, if you feel like it, which would be a good thing for you especially with how it echoes there. You'll feel good, you'll be surprised. Once, and I'll come. Understand me?"

Harry's mother holds Mrs. Dickstein's hand and squeezes it, and the two small women wheel on their fat legs like a couple of ducks and go to the door. Mrs. Dickstein, who is smaller even than Harry's mother, reaches to hug her. She makes motions with her lips, low cooing sounds, pats her shoulder, leaves. Harry's mother stands at the closed door and looks through its peephole out into the hallway. Then she turns around, her hand still on the doorknob. "She is coarse and she never finished high school. She's one of the finest people I know. I hurt her feelings by forcing her to leave."

We see Harry pushing the rocker back and forth, slowly. As he rocks it he says "You didn't have to make her leave for me if she's your friend. She worries about you, I'm glad she's here."

"So you don't have to worry?"

"I worry all the time."

"About *me?*"

"I think about you."

"Yes."

"It doesn't matter if you believe me. I don't know what it's supposed to mean to you anyway. I don't know what it means to me." He puts a cigarette in his mouth and lights it, blows out smoke, he sighs.

His mother says "Don't smoke, please." She stares at him.

Harry whines, nearly. "*She* did." Then, putting the cigarette, still lit and smoking, into an ash tray on the glass-topped table—his hand descending meets his hand coming up—he says "I guess you're selective about protecting yourself from pollution."

She purses her lips as if to say that she's waited for this. "Besides smoking cigarettes, which are disgusting and dangerous, she has television. Do you want her to give it up? I don't. She's lost enough. She's lost nearly everything."

"And I *have* everything?"

"I don't know what you have, Harry." She looks at the smoke coming up from the table and says "Do you mind doing that better?" He crushes the cigarette out.

He says "So why did you make her leave? We're finished."

"We're finished" his mother says. She takes her hand from the doorknob and it becomes a fist which she places against her stomach beside the other fist. "I wanted her to leave because—I thought this would happen. You're sentimental. You always complained about feeling guilty. You always said I made you feel guilty. I knew that I'd hear from you."

"So you disconnected the phone to make it difficult?"

"I don't know."

"To make it a kind of obstacle course I would get through if I were a worthy person. And then you could tell me so I'd know that you knew what I would do, like a good mouse in the laboratory, just in case I was feeling like a grown-up human being."

Her face is sealed in its wrinkles and clench. "You're a grown-up human being."

"Thank you."

We are above them again. From where we are we see no

face on Harry's body, only the top of his head and his moving shoulders and hands as he rocks the empty chair. "Do you know what day it is next week?" his mother says.

"My birthday."

"Do you know what that day means to me?"

"You mean now?"

"Do you know what it means?"

"I don't want you to tell me."

"I'm sure you don't. Don't worry, I'm not going to. I'm not going to. I'm not going to cry any more. But next week, I want you to know that I love you."

"That's meaningless, it isn't true."

"When they brought you to me in the hospital, do you know what I did? I cried."

"Could we not do this? It isn't fair—"

"I don't *care!* I have a right to say this! I cried because I knew you would leave me. I never thought you'd stay. But then, when you were a little child, I started forgetting what I had known when you were just born. That's why I got Belle to leave."

Harry is rocking the chair back and forth, he says nothing, we see no face.

His mother says "I thought—after all this, after all this time, *every*thing, I thought maybe now it isn't all dead. Maybe we can be friends, that's what parents and children should be. After everything, still, maybe friends."

Harry's "No" bangs against the sound of her last word. "No" he says. "Because you said that. Because of what you told me. It's a kind of tyranny you've always used, and I'm scared of it. I can't—"

She moves away from the door, comes toward him. He circles away from the chair and walks to the door. "Just because you're frightened" she says. "Everyone's frightened."

He is standing at the door and she is near the table with the glass top. He says "I have to go."

"You see how stupid I am" his mother says. "I never learned."

"Please don't tell me any more" Harry says. We cannot see his mother's face, only the top of her head. Harry's face is wrinkled because he is crying. "I'm going now" he says. "I have to go."

She follows him to the doorway. We see her small still back, his twisted face. The tone of her voice when she says "Good-by" is hard and certain, she is riding on the triumph of her tragedy.

"Good-by" he says.

Then, as if she were taller than he, looking down, she says "Call me by my name."

He opens the door, and we are in the outside corridor along the wall, waiting for him to come out. He says "Good-by, Mom?"

She says "My name. Use my *name!*"

His voice echoes in the hallway: "Claire?"

Her voice echoes too. "Yes."

"Good-by, Claire" he says.

And then he is in the hallway, she's pulled the door closed. There is a tinny popping sound from the far end of the hall as Mrs. Dickstein opens her peephole. Harry waits outside his mother's door, as if she will call, as if he will hear her fall to the floor of her apartment, or wail a final word, as if he will knock at the door. We are as close upon him as an ordinary lens might come without blurring the image. We draw back. So does he. He walks to the top of the flight of marble stairs, puts a hand on the banister, then starts to lower himself, pulling against his hand, a step at a time, as if his flesh has great weight. We are rising now as he sinks. His rate of descent is our rate of climb, through ceiling and floor and ceiling and floor and ceiling, and when he at last is outside on the hot silent street we crouch here on the air above him. His life is as small as a landscape now, his history is rolled out onto the earth, it is a map among other maps. From where he stands, the world is flat and endless. Here, above, we see that it is round, that it is two-dimensional, a bright picture of a bright shiny coin.

DATA

It had rained live tadpoles someplace in Germany beginning with an *S,* and in some other foreign place fresh blood had poured to the cobbled streets—this, according to *The Book of Fascinating Facts*—and now, most mornings, he checked outside the bedroom window as soon as he woke, looking (just in case) for something splendid and dreadful that dropped upon the earth.

His parents had given him *The Book of Fascinating Facts* when he was eight and he had turned it against them at once. He had told them that, actually, no, he *didn't* eat more cake than anyone: in 1788 a Swiss curd puller named Thrumm had eaten more cake than anyone before or since. And, in fact, there *wasn't* any reason not to swim right after eating: according to *The Book* it made no difference at all. And he had grown and they had died and *The Guinness Book of World Records* had thumped into ascendancy, and still—checking outside the bedroom window, or drinking what felt like a record amount of wine, or slipping on some slimy liquid in the streets—he leaned for a measure of amazement on *The Book of Fascinating Facts.*

He did it now, waking to the darkness of a nearly autumn morning in the hills. He went from the bed to the window and moved aside the curtains: the gray collapsing barn, the square small jungle still fenced in that had never been a garden, the trim banality of lawn around it (anyone could mow), the pine tree filled with sap bugs that, at three or four in the afternoon, lay against the back screen door unmoving, threatening in their very stillness; the corn behind the barn, the hayfield farther away than that, and then the highway and the hills and low puffy clouds that looked too heavy with darkness to be blown. The sky was clear of blood, amphibians, delight. A dozen buzzing barn swallows sat in their chubbiness on the wires that ran from the barn to the house. Most mornings he watched them, and they saw him most mornings and didn't show him if they cared.

Q: Do Animals Care?

A: Animals do not "care" as we humans understand that word! Animals have no feelings! They act automatically, in response to certain hereditary patterns! ("Hereditary" means ways of acting that are passed along from parents to offspring.) Sometimes we humans see patterns, like caring for the young, as evidence of emotions. This is incorrect! Yet— don't pets "love" their masters? Science has not found all the answers!

He took his clothing from the left side of the closet; nothing hung on the right. He ate his untoasted bread and drank his instant coffee at the kitchen table near the door. And when he went out along the path past the garden patch and sat in the car, he saw the tissues someone had left on the floor. Jammed between the seat rest and the cushion were the wrappers of hamburgers-to-go. He said "Sloppy goddam son of a bitch" and stalled the car.

A barn swallow dropped to the window of the car and lay against it, feathers spread, its head turned sideways as a person's would lie if he were listening. But this was a bird, it was watching, and the wet eye pulsed. He turned the key again and the engine raggedly nagged again and again and would not start. He stopped and turned it again and the starter-motor chirred in a lower key. He pumped the gas and turned the engine on and it tried not to cough. He looked at the swallow lying before him, staring, and the flesh of his face fell in upon his teeth, he tried to cry, he turned the key and heard a click like stones. He gripped the wheel and closed his eyes and screamed a high soft screech that had no energy behind it, that fell to the seat beside him and splashed.

The thin high cry awoke him and he said "Thank God. Thank God. Thank God. Thank God."

    Q: Can Dreams Come True?

    A: Apparently they can! Some people seem to experience things before they happen! Scientists cannot explain this phenomenon! (A "phenomenon" is something that is observed to actually happen. It often cannot be explained.) For example, there is the case of Ruben Jeffries of Southampton, England. In 1916, dreaming that something terrible had happened to his cousin in Galway, Ireland, Jeffries awakened in the middle of the night and sent a telegram to his cousin which said "How is your wife?" Back came the reply: "Shot by mistake outside our house. How did you know?" Jeffries had not seen or heard from his cousin for years! Such events are more common than we think, and no one has ever completely explained them!

He went to the window and saw that the office buildings across the courtyard were dark. What sky he could see above

Manhattan was blue with orange leaking across it in a ragged stain. He washed and then dressed from his suitcase, and then he stood so as not to wrinkle his suit pants. His clothing, made for someone fat, hung loosely. He walked in his room and then stood some more, pulling at the trouser crease. He called the hotel desk, being careful not to look at the mirror behind the phone, and they told him it was five o'clock in the morning. He delicately took his trousers off and, in shoes and socks and starched blue shirt, lay on the narrow bed, looking.

Q: How Quick Is "The Wink of An Eye"?

A: Possibly the quickest wink is that of the bald eagle! Its nictitating ("nictate" means to wink) membrane slides across its eye every 1/50 of a second! Not even the fastest, strongest blinker-of-a-human can match that!

At seven-thirty Washington Square Park was empty except for the sleeping drunks and a small blonde girl in khaki nodding on a distant green wood bench. The fountain was off, Garibaldi was on his horse forever, the Department of Sanitation truck drove past the park, washing a filth from the center of the street to its curbing, and the chlorine spray stayed in the air so that it felt for a breath as if the park were at the edge of a sea. The imitation Arc de Triomphe said all over FUCK and SUCK and PEACE and SANTORO 115. The sky was a lighter dark blue. Sounds of traffic in distress surrounded the park invisibly, and tired night custodians left New York University, across the street, as if they were cars.

He knew that Mrs. Miriam had owned her brownstone house—James had written *Washington Square* about a building four doors down—long after neighboring owners had sold to the school. When her daughter married for the third time Mrs. Miriam had sold the bottom two floors—perhaps in disgust with her past that had failed her, or because her daughter's alimony stopped. So he went through a university hollow-wood door with high black letters on it to enter her house, and he walked to the back of the building on yellow linoleum past letters on the yellow wall that told the students where to go for certain forms. But once up the metal stairs at the back, two flights, official ones, he came to wooden steps that moaned with age, and at the top of these he faced a door whose knocker was a gargoyle made of pitted brass. It had turned green, but didn't look old; it looked as if the gargoyle had caught a disease.

And then he was in the giant room with Mrs. Miriam, so small she could stand still and like a fawn in the woods be lost to sight. Her hair was white, her skin was whiter, the denim dress she wore came down to her shoes. She floated on her polished parquet floor, past littered tables whose legs were shaped in bells and swollen curves, around the flap-eared fireside chairs, between two blackened pianos, closed, among the many corrugated boxes someone had taped shut.

And he was drinking tea with Mrs. Miriam, who sat across from him in her Morris chair—her feet could not touch the floor—and looked at him and told him who he was: "It shocked me everytime I saw you, Harry. When I saw you first you weren't a year old and I was stunned how *big* you were. And when you were two and I saw you, or five, or eight—or when you graduated from school. My. You were always so *big*. That's how you look to me now, you know."

"Very large?"

"Yes. Do I look small?"

"You never were six feet tall, Mrs. Miriam."

"My. I never was *five* feet tall, my friend. And now I've grown less."

"Less?"

"Oh, a good deal less. Do you like the tea?"

"Delicious. Thank you."

"Honey."

"Excuse me?"

"I put buckwheat honey in it."

"Very delicious. Less?"

"What, Harry?"

"You say you've grown less?"

"Oh, considerably. Inches, several inches. I'm eighty-one years old, Harry. Some time ago I started to shrink."

"Oh. Do you—ah—"

"Yes, dear?"

"Feel it?"

"The shrinking?"

"Uhm."

"Did you ever feel yourself grow?"

"Ah."

She smiled and shook her head and rubbed together her fingers that had held the handle of her paper supermarket hot-drink cup. Then she held her hand in the air—he thought he saw the light come through it as if it were made out of fine porcelain—and she shook her head and said "I press them sometimes, or squeeze something for a little while, not very long even, and the skin stays flat. See that? All bloodless and flat? It takes so *long* for the blood to come back in." She lowered the hand a little and he stood and stepped to her chair and took her fingers in his fist and gently squeezed and released, squeezed and released. She closed her eyes, then opened them and said "Thank you, Harry, that will be fine," and he sat down again. She said "What a sweet surprise."

"Well, I hope they feel better soon."

"No, I didn't mean that. They won't. That's simply how they are now, you see. That is, the body simply *gets* that way. It doesn't go away. It isn't a matter of getting better, really. At this age it's simply what you've become. No, though, I meant that I'm surprised to see you, it's a pleasant surprise to see you in New York."

"Well, business—"

"A business trip to New York? I thought you worked in the country someplace, Harry. Upstate? Far away?"

"Syracuse. Outside of Syracuse."

"Yes, I remember that. And your business takes you to New York?"

"Not really, Mrs. Miriam."

"Oh?" Her face settled inward, all the little lines composed a picture of suspension. "You haven't any troubles to carry all that distance to me?"

"No, Mrs. Miriam."

"So you *didn't* come just to see me."

"Well, yes I did."

"You've come to see me?"

"I came to see—"

"But you didn't come for so many years, Harry."

"No."

"You know my daughter's dead. A little older than you."

"I'm sorry, Mrs. Miriam."

"She killed herself."

"I'm very sorry."

"Yes. She took too many sleeping pills. She was an alcoholic like her father."

"I'm very sorry, Mrs. Miriam."

"He killed himself the same way."

"I didn't know that."

"Yes. And your parents are gone, and that leaves me and a cousin in Hawaii, and heaven knows *she* hasn't long. She weighs three hundred pounds."

"And me."

"Pardon?"

"You and your cousin and me."

"But all those years I never *saw* you, Harry."

"Well I'm here, Mrs. Miriam."

"We're not responsible to one another now, Harry. Not after all this time. Surely—"

"I've always thought of you with such affection!"

Her face clenched closer. "Yes, but we've been separated by so many dyings. Haven't we?"

"Yes."

"Would you like some bread with your tea? All I have is bread. And margarine, if you care for that."

He said "I haven't come to make you responsible for me, Mrs. Miriam."

She smiled and set her paper cup on the floor that shone like light reflected from an eye.

He said "I'd like to ask you something, is all. I called and came over—is it too early, Mrs. Miriam? Shall I come back—"

"I don't protect my sleep very jealously" she said. "When it comes, when it decides to have me, then I sleep. Don't worry. And I wouldn't turn you out for an hour of sleep, Harry. You knew me so long—would I?"

He said "Thank you, Mrs. Miriam. I didn't sleep either. Not very much."

"It's such a difficult city to sleep in now."

"Thank you" he said. "Yes. Yes. I came here to ask you: do you remember my life when I was young?"

She looked at him and rubbed her fingertips together and he stood to reach for her hand. She said "No, no, never mind. Thank you, Harry, I remember *my* life. When *I* was young. Why do you want me remembering *yours?*"

He put his feet together and said "Because I can't."

"You can't remember your life?"

"I'm getting older" he said.

She said "I'm more than forty years older than you are, and so much further away from—and you think that *I* can remember everything you miss?"

"That's it!" He snapped his fingers and rubbed his hands together, nodded his head. He said "That's it! You see: I keep on *missing* it. This terrible nostalgia for—something. What? For *what?* I mean, I remember some things from my life with my parents—some of the arguments, some Christmases, giant boxes of gifts and my father smiling. You know, things like that. My mother taking me all the way to the shops and buying me a book of Jack London stories and a box of dominoes and me giving them back because I hated Jack London and dominoes more than anything else just then. And the dog my mother was scared of. My father in a straw hat coming home from Brooklyn College in the summer. You understand: I have some of that. But very little. I mean, for so many years—very little to keep. Do you see what I mean?"

"And you think because your mother and I were friends—"

"She lived down here in the Village when she was just out of school. They both lived here, in her apartment, after they got married. I mean, you saw them a lot, didn't you?"

"Yes. I knew your parents very well."

"And after they had me and moved, you still knew them."

"Yes. For some time."

"So maybe you remember some things—"

"It isn't likely, Harry."

"—that you could tell me about my life?"

"It isn't likely, Harry. No."

"You don't remember?"

"No. I mean, no, that isn't it. Would you like some tea?"

"Please. I mean, no thank you. No. Please: couldn't you tell me something?"

"What, Harry? What would you really like to know? You see, that's what I think I can't tell you: what you want to know."

"What do you think I want to know that's so mysterious, Mrs. Miriam? Why?"

"Suppose you tell me what you want to know."

"Couldn't you—"

"I knew Eugene O'Neill. Over there on that table, under those books over on the far left side? I have a whole folder of letters from him. I never answered one. Dreadful correspondent. I knew his wife. She wrote me two letters, I think, maybe only one. I never answered her either. Oh my."

"But couldn't you just kind of tell me *stories?*"

"Stories?"

"Couldn't you say what you remember and I could see if—"

"Tell you stories, Harry? At your age?"

"Mrs. Miriam, I have so *little* now!"

She looked at him and her small feet moved beneath her dress as if they were stirred by a current of air. The light outside the ten-foot windows far across the shiny room was gray and golden. The smears and dust and spider webs on the windows gleamed as if the light were thrown by them.

> Q: Are Spiders Insects?
>
> A: No! Despite common belief, spiders are *arachnids!* This is a different class of creatures from the insects! Spiders have eight legs!

Mrs. Miriam said "Harry, I remember some things that happened. Some ordinary events. I know some facts about your life, just as you must know a few about mine."
"Yes. That's what I want."
"Really?"
"That's what I want, Mrs. Miriam."
"I'm packing some of my things away, you see."
"Yes, Mrs. Miriam."
"When the boy from the market comes he brings up a few paper cartons for me, and I put away some of my books and letters and, oh, whatever comes to hand, really. Goodness. Afghans, I think it was. That was the last batch of belongings I packed."
"Yes. Are you moving, Mrs. Miriam?"
"Yes."
"Where to?"
"Don't you really know?"
"Well."
"Yes. That's right."
"Oh."
"Yes. You see there's no one left who loved my life to pack it up for me."
"Mrs. Miriam."
"And I don't want it left in a mess. Then scattered about by some workmen from that elephant of a university across the street where there are one or two men who actually wait every day, I think, to hear that I've finally died so they can move a bursar's assistant in here."

"Mrs. Miriam?"

"And I want it put in some order someplace. Away. I want it put away and saved. But out of sight. Away. Kept, but away. The way people put birth certificates in their safety deposit box: to save and never look at."

"But wouldn't you rather look at what you were?"

"Harry, don't make stupid small talk! I *am* what I was!"

"I don't know what I was, Mrs. Miriam!"

"Do you have mirrors in your house? Then look. That's what you were. Period. With your mother and your father and the house you all lived in, the clothes they used to wear. It's done and put away and you should leave it. That's the only story I can tell you, Harry. Leave it where it is."

"But what is *it?* What do you mean? What?"

Q: How Large Is the Human Body?

A: Larger than you think! The skin of an average adult equals 18 square feet when laid out flat! There are 100 sweat glands in one square inch of human skin! A person's veins, arteries and capillaries, if laid out end to end, could go around the world 2½ times!

So he woke in the darkness of the upstate hills and went from his bed to the window. The fat unfrightened swallows ticked and buzzed at the dawn. He turned his back to them and looked across the room, past the bed, to the wide bureau with its hinged high mirror, and debris—some pennies and a nickel and a dime, his wallet, a book of matches, keys, his glasses. The mirror was tilted to capture the bed, which looked like the scene of a fight. He looked at the mirror, which couldn't see him, and dressed himself at the far side of the room, then went down. In the trousers he'd worn from New York, in the same socks and underclothes, he made his coffee and ate his bread and went outside to the car. Above him the birds cried, took off and dove and returned to their wire and watched him sideways, as the mirror had. He hunched his shoulders and looked beyond the swallows to the skies, to where the frogs and blood would fall from if ever the secret facts of his life came true, dropping from the heavens, turned from words to flesh. His engine started. He engaged the gear lever, and the car began to slowly roll in reverse. He put his hands on the wheel and backed down the driveway, then he drove himself away.